D1194064

Biochemical Fluorescence: Concepts

VOLUME 1

Biochemical Fluorescence: Concepts

VOLUME 1

EDITED BY

Raymond F. Chen
Laboratory of Technical Development
National Heart and Lung Institute
National Institutes of Health
Bethesda, Maryland

AND

Harold Edelhoch
Clinical Endocrinology Branch
National Institute of Arthritis,
Metabolism, and Digestive Diseases
National Institutes of Health
Bethesda, Maryland

Marcel Dekker, Inc.   New York

PREFACE

In the 1960's, fluorescence spectroscopy became an important
tool for the physical biochemist. Much of the reason for this pop-
ularity was due to the sensitivity and versatility of fluorescence
methods as well as to the pioneering theoretical and practical work
by investigators such as Förster and Weber. The last decade has
seen an expanding interest in luminescence methods; indeed, we appear to be in a period where the literature on fluorescence is
growing exponentially.

Luminescence phenomena involve ground and excited states, and
the study of such phenomena may require different kinds of measurements, such as fluorescence yield, decay, polarization and spectral
distribution. While the complexity of luminescence should not dis-
courage investigators from using the technique, it is clear that
this is really not a single method but a collection of methods, each
of which has been developing as a result of the wide range of appli-
cability of luminescence spectroscopy.

There is obviously no point in trying to assemble a "text" on
fluorescence when the field is changing so rapidly, but there does
seem to be a need to recapitulate some of the concepts and applications of fluorescence on which future advances will be based. Much
understanding has already accumulated about the basis of fluorescence, polarization, energy transfer, quenching mechanisms, and
spectral shifts, and has given rise to many valuable studies on specific biological systems. We now see further technical innovations
brought about by the advent of the digital computer, the laser, and
new electronic developments. A glance through the table of contents
of this volume indicates that the area of decay kinetics has been
opened up by technical advances, especially time-correlated single

iii

photon methods and the associated computational theory. Other new
techniques include circular polarization of luminescence, stopped-
flow fluorescence, fluorescence-monitored chemical relaxation, and
the evaluation of relative orientation by polarized excitation energy transfer.

Volume 2 will deal with some of the newer applications of fluorescence spectroscopy. New fluorescent probes and quenchers have
helped to open up areas such as membranes, muscle and nerve components, and other subcellular organelles.

Raymond F. Chen
Harold Edelhoch

National Institutes of Health
Bethesda, Maryland

CONTRIBUTORS TO VOLUME 1

Robert E. Dale, Mergenthaler Laboratory for Biology and McCollum-Pratt Institute, The Johns Hopkins University, Baltimore, Maryland

Josef Eisinger, Biophysics Department, Bell Laboratories, Murray Hill, New Jersey

Thomas M. Jovin, Department of Molecular Biology, Max Planck Institut fuer Biophysikalische Chemie, Goettingen, German Federal Republic

Irvin Isenberg, Department of Biochemistry and Biophysics, Oregon State University, Corvallis, Oregon

Stuart A. Levison, Department of Biochemistry, Scripps Clinic and Research Foundation, La Jolla, California

Peter W. Schiller, Clinical Research Institute of Montreal, Montreal, Quebec, Canada

Izchak Z. Steinberg, Department of Chemical Physics, The Weizmann Institute of Science, Rehovot, Israel

Philippe Wahl, Centre de Biophysique Moléculaire - C. N. R. S., Orléans, France

CONTENTS

Cumulative Indexes will appear at the end of Volume 2.

CONTENTS OF VOLUME 2

Biochemical Fluorescence:
Concepts

VOLUME 1

Chapter 1

DECAY OF FLUORESCENCE ANISOTROPY

Philippe Wahl
Centre de Biophysique Moléculaire — C.N.R.S.
Orléans, France

I. INTRODUCTION

The fluorescence polarization of dye solutions was discovered by Weigert [1]. He found that polarization changes with the solvent viscosity. This was an indication that the Brownian motion is an important factor which affects the polarization.

The theory of the Brownian depolarization was given by F. Perrin [2,3] for rigid molecules of spherical and ellipsoidal shape. In In this theory, mathematical expressions are given which relate the degree of polarization to the lifetime of the dye, to the molecular volume (and to the axial ratio for ellipsoidal molecules), as well as to the temperature and the viscosity of the solvent [4,5,6]. One now prefers to use the emission anisotropy, introduced by Jablonsky [7], instead of the degree of polarization. We will use the emission anisotropy factor in this article.

The theory was first tested by Perrin [3] with chromophore solutions having different viscosities. These experiments were done with a steady excitation, and the quantities measured are what we shall call the static parameters. Perrin took the volume found by viscosity measurements [8] which enabled him to determine the lifetime of fluorescein which was in good agreement with the direct determination of Gaviola [9]. Perrin's theory, however, does not describe satisfactorily all the aspects of the experiments performed with aromatic molecules [10,11]. One of the reasons for this discrepancy is probably that these molecules are too small to obey the law of rotational Brownian motion.

On the basis of Perrin's theory, G. Weber [12,13] proposed a method for the determination of the morphological parameters of macromolecules in solution. In this method a relaxation time is obtained by measuring the static emission anisotropy at different temperatures, or in a series of solvents having different viscosities. This method has been applied to a variety of proteins, nucleic acids, and membranes labeled with fluorescent chromophores. At first sight, one may think that macromolecules obey the laws of Brownian motion fairly well (see reviews of Chen et al. [14] and Dandliker et al. [15]). A critical analysis, however (Wahl [16], Wahl and Weber [17]), shows that, in many cases, the variation of the temperature or of the nature of the solvent may influence the anisotropy in a complicated way. Therefore a simple application of Perrin's formula is not justified and might even lead to a wrong interpretation of the experiments [18].

Perrin's theory predicts that the decay of the principal polarized fluorescence components $I_{\parallel}$ and $I_{\perp}$ are influenced by the Brownian motion. As Jablonski [7,19,20,21] has pointed out, the average lifetimes measured with a phase fluorometer are different for the whole fluorescence and for each of the polarized components. Coupling lifetime measurements in polarized light with measurements of static anisotropy should permit the determination of the fundamental anisotropy and the volume of a spherical chromophore. These data are obtained in a given solvent and at a given temperature. Experiments performed by Szymanowski [22], Kessel [23], and recently by Bauer [24] are in fair agreement with the prediction of Jablonski.

Spencer and Weber [25] have discussed in detail the use of the phase fluorometer to obtain the Brownian correlation times and the influence of the modulation frequency. Most of the present phase fluorometers work at one or two modulation frequencies [26]. The method might gain in value if the measurements were made in a large continuous range of frequency. In the present state of the technique, the phase method is in many respects surpassed by the pulse methods. Among these, the single photoelectron counting method presents a set of advantages not found in other methods [27,28,29, 29a]. It is often possible with this method to follow the time course of the emission anisotropy during an appreciable fraction of the fluorescence decay and is currently being applied in the determination of correlation times in a number of biological compounds [29a, 30, 30a]. The method may also be applied to the study of energy migration between a set of identical chromophores [31,32].

It is well known that energy transfer between identical molecules decrease the static anisotropy [33,34,35]. Consequently, the decay of anisotropy obtained after a flash excitation may be influenced by energy migration. One must be aware of this possibility when interpreting such measurements, and it is necessary to separate Brownian and transfer contributions [31]. A quantitative analysis of the transfer contribution may bring about detailed information on the spatial distribution of the chromophores. This principle has been applied to the study of the ethidium-DNA complex [32,36,36a].

I shall first review the principle of the experimental determination of anisotropy decay. Then the theory of Brownian depolarization will be discussed and its application to the study of macromolecules. Finally, some aspects of energy migration will be given.

II. EXPERIMENTAL DEFINITION OF THE EMISSION ANISOTROPY

A. Definitions

The state of the fluorescence polarization may be characterized by the three polarized components Ix, Iy, Iz. Their sum

$$S = Ix + Iy + Iz \tag{1}$$

is proportional to the fluorescence flux radiated in all directions. The excitation is due to the local electric field propagated by the exciting light. Then, if the exciting light is polarized linearly along OZ, one has by symmetry

$$Ix = Iy \tag{2}$$

If the exciting beam is made of natural light vibrating in the XOZ plane, then

$$Iz = Ix \tag{3}$$

By observing the fluorescence along OX, one sees

$$I_\parallel = Iz \qquad I_\perp = Iy \tag{4}$$

The anisotropy of emission [7] may be defined as

$$r = D/S \tag{5}$$

with

$$D = I_\parallel - I_\perp \tag{6}$$

According to Eqs. (1-5), the emission anisotropy for a vertically polarized exciting light is

$$r = \frac{I_\parallel - I_\perp}{I_\parallel + 2I_\perp} \tag{7}$$

and for a natural exciting light

$$r_n = \frac{I_\parallel - I_\perp}{2I_\parallel + I_\perp} \tag{8}$$

One may easily show that between the emission anisotropy r and r_n one has the relation

$$r = 2r_n$$

The use of emission anisotropy instead of the degree of polarization leads to simpler theoretical interpretations of the experiments. These parameters are related to each other by simple expressions [7].

The emission anisotropy of a solution containing several molecular species is given by the following expression [7,12]:

$$r = \Sigma r_k f_k \tag{9}$$

where r_k is the emission anisotropy of the species k, and $f_k = S_k/S$ is the fractional fluorescence intensity of the species k. All these definitions and expressions are valid for continuous excitation as well as for an excitation by an infinitely short flash.

B. Anisotropy Decay

If the fluorescence is excited by an infinitely short flash, the intensities $I_\parallel$, $I_\perp$, S, and D become time dependent. S is proportional to the fluorescence decay and, in the simplest cases, decays as a single exponential:

$$S = S_0 e^{-t/\tau} \tag{10}$$

where τ is the lifetime of the excited state.

The anisotropy decay will be defined as

$$r(t) = D(t)/S(t) \tag{11}$$

These functions are not directly measurable in a pulse fluorometer because the exciting flash is not infinitely short. One directly measures the experimental functions $i_\parallel(t)$ and $i_\perp(t)$. Let us assume that the exciting light is vertically polarized. Then one calculates [37,38]

$$d(t) = i_\parallel(t) - i_\perp(t)$$
$$s(t) = i_\parallel(t) + 2i_\perp(t) \tag{12}$$

The functions D(t) and S(t) are related to these experimental functions by the following convolution integrals:

$$d(t) = \int_0^t D(t - T)g(T)\, dT$$

$$s(t) = \int_0^t S(t - T)g(T)\, dT \tag{13}$$

where g(T) is the response function of the pulse fluorometer. This function essentially depends on the excitation function (time distribution of the intensity in the flash) and on the response function of the photomultiplier. The determination of this function is discussed elsewhere [27,28,29,38a]. The resolution of Eq. (13) is

difficult. Several methods have been proposed [28,39,40,40a,40b].
In the special case of measurements in polarized light, we generally
use the following procedure: First, we try to determine S(t) by one
of the methods proposed; we always check the result by the synthetic
method. We put:

$$D(t) = r(t)S(t) \tag{14}$$

where S(t) is the function previously determined; we then try several r(t) functions in order that the numerically computed convolution d(t) may fit the experimental curve. A specially simple case
occurs when S(t) and r(t) are single exponentials:

$$S(t) = S_0 e^{-t/\tau}$$

$$r(t) = r_0 e^{-t/\theta}$$

In this case, D(t) is also a single exponential

$$D(t) = D_0 e^{-t/\tau'}$$

and one obtains the relation

$$\theta = \frac{\tau\tau'}{\tau - \tau'}$$

and   (15)

$$r_0 = \frac{D_0}{S_0}$$

C. Relation Between the
Static Anisotropy and the Anisotropy Decay

A continuous light source may be considered as a sum of an
infinite number of flashes [3]. As a result, one obtains for the
static anisotropy

$$\overline{r} = \frac{\int_0^\infty D(t)\,dt}{\int_0^\infty S(t)\,dt} = \frac{\int_0^\infty d(t)\,dt}{\int_0^\infty s(t)\,dt} \tag{16}$$

If $S(t)$ is a single exponential like (10), Eq. (16) gives

$$\overline{r} = \frac{1}{\tau} \int_0^\infty r(t) e^{-t/\tau} \tag{17}$$

Put the case that $S(t)$ and $D(t)$ are single exponentials; Eq. (17) leads to:

$$r = r_0 \frac{\tau'}{\tau} = \frac{r_0}{1 + \tau/\theta} \tag{17a}$$

The last member is equivalent to the well-known Perrin's formula.

D. Experimental Details of the Anisotropy Measurements

One must first measure the decay of the principal·polarized components of the fluorescence $i_{\parallel}(t)$ and $i_{\perp}(t)$. The method used is generally the single photoelectron counting method [41,42]. The general principles of this method are given in several review articles [27,28,29,29a] and special setups for polarization measurements have been described [28,29a,30,43-46].

The gas contained in the flash lamp must be chosen to give the wavelengths wanted for the excitation. The selection of the wavelength band is made with an optical filter or a monochromator. The light of the exciting beam may be vertically polarized by a polarizer, or completely unpolarized. At the output of a monochromator, a Lyot depolarizer is necessary to obtain nonpolarized light [47].

The cuvet which contains the sample must be thermostated. The fluorescence light is filtered by an optical filter in order to eliminate the scattered or reflected light. Stray light is generally strongly polarized and may cause an error at the beginning of the $d(t)$ curve. The two components $i_\parallel$ and $i_\perp$ are measured with the same photomultiplier. A rotating polarizer (polaroid, for instance) successively selects one of the components. For accurate measurements, one is led to store the information in the multichannel analyzer for a long time. In order to eliminate the influence of the intensity and frequency fluctuations of the exciting flash, $i_\parallel$ and $i_\perp$ are alternately measured many times. It is convenient to store each component in a distinct section of the analyzer memory. A device has been contrived to automatically perform these operations [28].

In case this automatic device should not be available, the intensities $i_\parallel$ and $i_\perp$ may not be in the ratio which allows a correct determination of $s(t)$ and $d(t)$. It is then necessary to correct one of the polarized components, say $i_\perp$. This can be done as follows:

For several short periods of time during which the analyzing polarizer is alternately set in its two positions, the sums of impulsions $N_\parallel$ and $N_\perp$ are measured at the "time-to-height converter" output. $i_\perp$ is corrected by the multiplicative factor

$$\frac{\int_0^T i_\parallel(t) \, dt}{\int_0^T i_\perp(t) \, dt} \times \frac{N_\perp}{N_\parallel}$$

where T represents the measurement time range defined by the converter.

The time-to-amplitude converter may drift if the measurements last a long time. This may be corrected with a pic stabilizer working on signals from a reference photomultiplier viewing the flash directly [28,44]. The simultaneous storage of the two polarized components should accelerate the collection of the data [46]. But one must be sure that the two channels are identical, that the two

photomultipliers have the same impulse response function, and that
the difference in transit times is compensated.

In some multichannel analyzers one can directly compute $d(t)$
and $s(t)$ by a transfer operation from one memory section to another.
If this is not possible, one can record the data on paper tape and
perform the calculation on a computer. $s(t)$ may be directly measured if the polarizer placed in front of the photomultiplier makes
a given angle with the vertical [7,25]. With an excitation which is
vertically polarized, this angle is about 55 deg or 35 deg with unpolarized exciting light.

III.   THEORY OF FLUORESCENCE ANISOTROPY

A.   Transition Moments

According to the quantum theory of visible and ultraviolet molecular spectra, the absorption of a photon produces the transition
of a molecule from the ground state to a vibronic excited state
$(S_0 \to S_n)$. A transition moment with a well-defined direction in the
molecule is associated with such a vibronic transition. The probability of molecular excitation is proportional to the cosine square
between the corresponding moment and the electric vector of the exciting beam. After a certain time the excited molecule may emit a
photon of fluorescence which corresponds to a vibronic transition
from the lower excited state to a vibrationally excited ground state
$(S_1 \to S_0)$. This vibronic transition is characterized by a transition
moment which determines the polarization of the molecular emission.

The conditions of excitation which prevail in the spectroscopy
of solutions of organic molecules generally involve several vibronic
transitions. Several directions of transition moments may be involved in such cases, especially when the transition is electronically forbidden [48,49]. The directional properties of the excitation

probability of a single molecule may be characterized by a second-
order tensor [5,50]. In the same way, the emission of a single
molecule is also characterized by another second-order tensor. In
an ordinary solution the molecules are in their ground state and are
isotropically distributed. But after an excitation by light, the
excited molecules are anisotropically distributed, since the excitation probability has an angular dependence. This is called photoselection [48]. The fluorescence emission of a solution is the sum
of the emissions of the excited molecules. This emission is polarized since the excited molecules are anisotropically distributed.
Furthermore, all the events which modify this distribution during
the lifetime of the excited state influence the fluorescence polarization of the solution.

B. General Expression of the Emission
Anisotropy of a Solution of Identical Molecules

In a very general mathematical analysis, Soleillet [50] has
shown that the polarization properties of a fluorescent substance
may be characterized by a tensor of the fourth order. For an isotropic medium, like a solution, the state of polarization is completely described by the three invariant P, Q, R of what Soleillet
calls "the anisotropic fluorescent element isotropically distributed."

The emission isotropy of a solution, excited by a linearly polarized light, is given by the expression [51]

$$r = \frac{3(Q + R) - 2P}{10P} \qquad (18)$$

This general formula includes the case of circular polarization.
With aromatic chromophores the circular polarization of the fluorescence is negligible compared to the linear components [52]. In this
case, $Q \equiv R$. With a flash excitation, these quantities are time dependent, and if the solution contains identical molecules having
identical probabilities of emission, one obtains

$$P(t) \approx S(t)$$

$$Q(t) \approx q(t)S(t) \tag{19}$$

where $S(t)$ is the decay law of the whole fluorescence. Finally, the time-dependent anisotropy may be written

$$r(t) = \frac{3q(t) - 1}{5} \tag{20}$$

It is interesting to note that the decay of anisotropy as expressed by (20) does not depend on the fluorescence decay law but only on the phenomena which modify the orientation of the transition moments. This is true only when the decay law is identical for all the emitting molecules. We will now distinguish between several experimental situations. We first assume that the chromophores are widely dispersed in the solution. Excitation transfers are then impossible. Furthermore, we will distinguish between rigid and fluid solvents.

C. Rigid Dilute Solution

In a rigid solution, Brownian motion cannot occur. A molecule excited to a vibronic state falls to the lowest electronic excited state with a low vibrational energy. This process of internal conversion is very fast (a few psec). It can be considered to occur instantaneously before any fluorescent emission. Consequently, the transition moments associated with absorption will generally be different from the transition moment of emission. The orientation will be time independent, as will be the anisotropy. One may write

$$q(t) \equiv q(0) \equiv \overline{q} = \sum_{i,j} \epsilon_{ij} e_{ij} \qquad i,j = 1, 2, 3 \tag{21}$$

where q is the static quantity (corresponding to the continuous excitation). ϵ_{ij} and e_{ij} are the components of the molecular absorption tensor and the molecular emission tensor in molecular axes. If the absorption and the emission are strongly allowed electronic

transitions, and if the measurements are made in a wavelength domain
close to the 0-0 vibronic transition, there is a single transition
moment with a well-defined direction. If this is the case,

$$\epsilon_{ij} = \alpha_i \alpha_j$$
$$e_{ij} = \beta_i \beta_j \tag{22}$$

where the α_i's and β_i's are the director cosines of the transition
moments of absorption and emission. Then from (21) $q(0) = \cos^2 \lambda$
and from (20)

$$r_0 = \frac{3 \cos^2 \lambda - 1}{5} \tag{23}$$

r_0 is the fundamental anisotropy and λ the angle between the two
transition moments. Equation (23) predicts that

$$-0.2 < r_0 < 0.4 \tag{24}$$

The maximum (0.4) is obtained for $\lambda = 0$ and the minimum ($r_0 = -0.2$)
for $\lambda = 90$ deg.

In the molecules of symmetry, D_2, D_{2h}, and C_{2v}, the different
transition moments are oriented along three perpendicular axes [49].
Taking these axes as reference axes, one has

$$\epsilon_{ij} = e_{ij} = 0 \qquad \text{for } i = j,$$

and

$$\epsilon_{11} = a_x \qquad \epsilon_{22} = a_y \qquad \epsilon_{33} = a_z$$
$$e_{11} = e_x \qquad e_{22} = e_y \qquad e_{33} = e_z$$

Equations (20) and (21) give

$$r_0 = \frac{3(a_x e_x + a_y e_y + a_z e_z) - 1}{5} \tag{25}$$

with $\Sigma a_x = \Sigma e_x = 1$, where a_x is the fractional number of the molecules which are excited with a transition moment in the x direction,
and e_x is the fractional number of the molecules which emit with a

transition moment in the x direction [49]. r_0 depends on the wavelength of absorption and emission. It can be shown that its value
is always in the range determined by Eq. (24). The value 0.4 has
never been observed. Some strong fluorescent dyes in solid solution
may have a value very close to 0.4, but still smaller. Jablonski
[53] assumes that this effect is due to torsional vibration. If we
assume the same torsional angle ϵ in the ground state and in the
excited state, we get

$$r_0 = 0.4 \left(\overline{\frac{3 \cos^2 \epsilon - 1}{2}} \right)^2 \tag{26}$$

where the bar means an average over all the possible orientations
around the average position of the transition moment.

IV.   BROWNIAN DEPOLARIZATION

A.   Theory

In fluid solutions, molecules are submitted to the rotational
Brownian motion. During the time between excitation and emission,
molecules rotate randomly. The initial angular distribution, which
is created by a flash excitation at time zero, is progressively destroyed and the fluorescence anisotropy decays with the fluorescence
intensity. In this case one may write [5]

$$q(t) = \sum_{i,j} \epsilon_{ij} <E_{ij}(t)> \tag{27}$$

As previously, ϵ_{ij} are the components of the absorption tensor in
some molecular axes. Let us consider three laboratory axes $OX_0Y_0Z_0$
which coincide with the molecular axes of a given molecule at the
time of excitation. $E_{ij}(t)$ are the components of the emission tensor
in the $OX_0Y_0Z_0$ axes, at time t after the excitation. One has to
take the average of this quantity over all the possible orientations
resulting from the Brownian rotation. When the absorption and the

emission are each characterized by a single transition moment, Eq. (27) becomes

$$q(t) = <\cos^2 \Phi(t)> \tag{28}$$

where $\Phi(t)$ is the angle formed between the direction of the absorption transition moment at the time of excitation and the direction of the emission transition moment at the time of emission. A direct proof of this formula has been recently given [53a].

The rotational Brownian motion depends on the structure, size, and shape of the molecule; one will obtain different expressions, depending on the molecular structure, for the decay of anisotropy. Analytical calculations have been performed for different molecular models. Monte Carlo calculations have also been recently used to predict the course of the anisotropy decay [54].

1. Spherical Molecules

In this case, the rotational motion is isotropic. The application of Eqs. (23) and (28) gives [3,6]

$$r(t) = r_0 \frac{<3 \cos^2 \omega(t)> - 1}{2} \tag{29}$$

where $\omega(t)$ is the rotational angle of a spherical diameter between the excitation time and the emission time

$$\cos^2 \omega(t) = \int_0^\pi \cos^2 \omega \; W(\omega,t) \; \sin \omega \; d\omega \tag{30}$$

$W(\omega,t)$ is a probability density function which is a solution of the diffusion equation of a sphere diameter

$$\frac{\partial W(\omega,t)}{\partial t} = D\nabla^2 \; W(\omega,t)$$

with the initial condition $W(\omega,0) = \delta(0)$. D is the coefficient of rotational diffusion of a sphere

$$D = \frac{kT}{6\eta V}$$

where k is the Boltzmann constant, T the absolute temperature, η

the solvent viscosity, and V the hydrodynamic volume. One obtains
[21]

$$r = r_0 e^{-t/\theta} \tag{31}$$

where the correlation time θ is given by

$$\theta = \frac{1}{6D} = \frac{\eta V}{kT} \tag{32}$$

2. Ellipsoidal Molecules

The theory of the Brownian depolarization was first established
by F. Perrin [4,5]. This author used the tensor formulation of
Soleillet [50]. From the calculations of Perrin, it is easy to determine the decay of anisotropy of an ellipsoid. For simplicity,
we consider only the case of a transition moment without circularity.
Then Eqs. (20) and (27) are applicable. It is convenient to choose
as molecular axes the principal axes of the ellipsoid. One may write

$$<E_{ij}(t)> \; = \; \sum_{m,n} e_{mn} <C_{mi}(t)C_{nj}(t)>$$

$$m, \; n, \; i, \; j = 1, \; 2, \; 3 \tag{33}$$

where e_{mn} are the components of the emission tensor ·in the molecular
axes and $C_{mi}(t)$ and $C_{nj}(t)$ are the cosine directors of the molecule
axes at time t in the $OX_0 Y_0 Z_0$ axes. The value of $q(t)$ may be computed in the same manner as done by Perrin for the quantity corresponding to the static experiments. One obtains

$$q(t) = \sum_i A_i <C_{ii}^2(t)> \; + \; 2B_i <C_{jk}^2(t)>$$

$$+ \; 2Di <C_{jj}(t)C_{kk}(t) + C_{jk}(t)C_{kj}(t)> \tag{34}$$

where A_i, B_i, D_i are symmetrical functions of ϵ_{ij} [5]. The value of
the average function appearing in Eq. (34) has been computed by
Perrin who used the diffusion equation of ellipsoids [Eqs. (24),
(32), and (33) of Ref. [5]]. Finally, $q(t)$ and $r(t)$ may be expressed
as a sum of five exponential functions.

This result has recently been confirmed by several authors [55,
55a, 55b] when absorption and emission are each characterized by a

single transition moment. The calculation of Tao [43] is valid when
the two transition moments have the same direction.

The sum is reduced to three exponentials when the ellipsoid is
symmetrical [38,43,51]

$$r(t) = A_1 e^{-t/\theta_1} + A_2 e^{-t/\theta_2} + A_3 e^{-t/\theta_3} \tag{35}$$

where the three correlation times θ_1, θ_2, θ_3 are functions of the
rotational diffusion coefficients D_z and D_x characterizing Brownian
rotation around the axis of symmetry and around a transverse axis,
respectively:

$$\theta_1 = \frac{1}{2D_x + 4D_z}$$

$$\theta_2 = \frac{1}{5D_x + D_z} \tag{36}$$

$$\theta_3 = \frac{1}{6D_x}$$

D_x and D_z have been computed by Perrin as a function of the molecular volume V and the axial ratio. A tabulation of θ_1, θ_2, θ_3 is
given by Tao [43]. The three amplitudes A_1, A_2, A_3 are functions of
the intrinsic tensors of absorption and emission ϵ_{ij} and e_{ij}. For
the case of a single transition moment in emission and absorption,
one obtains

$$A_1 = 0.3 \sin^2 \delta_1 \sin^2 \delta_2 \cos^2 (\epsilon_1 - \epsilon_2)$$

$$A_2 = 0.3 \sin 2\delta_1 \sin 2\delta_2 \cos (\epsilon_1 - \epsilon_2) \tag{37}$$

$$A_3 = 0.1(3 \cos^2 \delta_1 - 1)(3 \cos^2 \delta_2 - 1)$$

where δ_1, ϵ_1 and δ_2, ϵ_2 are the polar coordinates of these two
transition moments [56] (see Fig. 1).

For the transition moments of absorption and emission which
are parallel and which oscillate rapidly around a position of equilibrium [51], one has

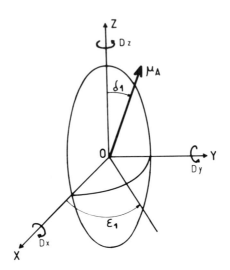

FIG. 1. Symmetrical ellipsoid: orientation of the transition moment μ_A is defined by the polar coordinates δ and ϵ relative to the principal axes of the ellipsoid.

$$A_1 = r_0 \frac{3}{4} \sin^4 \delta$$

$$A_2 = r_0 \frac{3}{4} \sin^2 2 \delta \qquad\qquad (38)$$

$$A_3 = r_0 \left(\frac{3 \cos^2 \delta - 1}{2} \right)^2$$

with

$$r_0 = \frac{(3 \langle \cos^2 \epsilon \rangle - 1)^2}{10}$$

where δ is the angle defining the average orientation of the transition moment with the axis of symmetry, and ϵ is the amplitude of oscillation around this average orientation.

With proteins labeled with fluorescent dyes, there is often some reason to assume that the dyes are randomly oriented around the molecular axes. In that case, A_1, A_2, A_3 take their average values, which are [18]

$$A_1 = A_2 = 0.4 r_0$$

$$A_3 = 0.2 r_0 \qquad\qquad (39)$$

3. Internal Rotation

The fluorescence emission of a macromolecule is often due to a small aromatic chromophore bound to the macromolecule backbone by a covalent link. Then there is a possibility for this chromophore to rotate around this bond, a motion which will influence the decay of fluorescence anisotropy. Calculations for a model concerning this situation have been performed by Gottlieb and Wahl [57]. In this model, a spherical chromophore rotates around its link which binds it to a spherical macromolecule. If the chromophore rotates freely around its link, the anisotropy is given by

$$r(t) = e^{-t/\theta} (A_1 e^{-2t/\theta_i} + A_2 e^{-t/6\theta_i} + A_3) \tag{40}$$

where θ and θ_i are the correlation times of the macromolecule and of the chromophore, respectively.

The expressions of A_1, A_2, and A_3 are still given by Eq. (37). Here the direction of the OZ axis is chosen along the bond which links the chromophore to the macromolecule. If the macromolecule is very big, the anisotropy is given by the term in parenthesis in Eq. (40). For t going to infinity, the anisotropy becomes equal to the constant term A_3. One may also assume that the chromophore performs jumps of fixed angles. Formulas have been computed for these cases also [57]. We simply give here the formula relative to three positions with angles of 120 deg between them. The frequency w of the jumps is assumed to be the same for all the jumps between neighboring positions. One obtains

$$r(t) = e^{-t/\theta} [(A_1 + A_2)e^{-3wt} + A_3] \tag{41}$$

Wallach [58] has generalized these calculations when several covalent links separate the chromophore from the macromolecule.

Another model has recently been envisaged for chromophore rotation about an axis. In this model the free rotation is limited to a given amplitude "ℓ" by two reflective barriers. Calculations were carried out in two extreme cases. In one case, each collision of the excited chromophore against a barrier is assumed to entail a

deactivation while in the second case the rate of deactivation is
not affected by these collisions [53a].

The expression of anisotropy corresponding to this last case
is

$$r(t) = A_3 + A_1 \frac{\sin^2 \ell}{\ell^2} + A_2 \frac{\sin^2 (\ell/2)}{(\ell/2)^2} + \sum_{n=1}^{\infty} [A_1 B(n,\ell)$$

$$+ A_2 C(n,\ell)] \exp[- \frac{n^2\pi^2}{\ell^2} Dt]$$

(41a)

where D is the chromophore coefficient of rotational diffusion.

In the sum of Eq. (41a), only a few terms have to be considered,
since the others are negligible. For $t = \infty$, Eq. (41a) becomes

$$r(\infty) = A_3 + A_1 \frac{\sin^2 \ell}{\ell^2} + A_2 \frac{\sin^2 (\ell/2)}{(\ell/2)^2}$$

When excitation occurs via the $S_0 \to S_1$ transition, A_1, A_2, and A_3
are positive. The asymptotic anisotropy $r(\infty)$ is then greater than
A_3 which is the asymptotic anisotropy corresponding to free rotation
[see Eq. (40)].

More generally, one expects that the anisotropy decay characterizing a limited local motion of a chromophore bound to a macromolecule or to a macroscopic structure such as a membrane is given by an
expression having the following form:

$$r(t) = r_0[af(t) + 1 - a]$$

(42)

where a is a constant given by

$$r_\infty = r_0(1 - a) = r_0 \frac{(3 \overline{\cos^2 \beta} - 1)}{2}$$

(43)

$\overline{\cos^2 \beta}$ is the average of all the possible angles between the absorption transition moment at the time of absorption and the emission
transition moment at the time of emission, $f(t)$ is a decaying function when excitation occurs via the $S_0 \to S_1$ transition.

4. Flexible Macromolecules

The fluorescence polarization has been calculated for the Rouse

model of the polymer chains [16]. On the other hand, the Monte
Carlo method has also been used to study this problem [59].

5. Initial Slope of the Anisotropy Decay

If one assumes that the anisotropy is a sum of exponentials,
one may write:

$$r(t) = r_0 (\sum_i a_i e^{-t/\theta_i})$$ (44)

with $\sum_i a_i = 1$. For a small value of t/θ_i, this function may be approximated by the following single exponential function [38,44]:

$$r = r_0 e^{-t/\theta_h}$$

where θ_h is a harmonic average defined by

$$\frac{1}{\theta_h} = \frac{1}{3} \sum \frac{a_i}{\theta_i}$$ (45)

For ellipsoid molecules bearing chromophores randomly oriented, it
has been shown [12,60] that

$$\frac{1}{\theta_h} = \frac{1}{\rho_x} + \frac{1}{\rho_y} + \frac{1}{\rho_z}$$ (46)

where ρ_x, ρ_y, ρ_z are the three principal rotational relaxation times
which characterize the dielectric absorption and which are related
to the rotational coefficient of diffusion by three relations of the
type

$$\rho_x = \frac{1}{D_y + D_z}$$

If the transition moments have a definite orientation in the molecule, θ_h is a weighted average which may be written

$$\frac{1}{\theta_k} = \frac{a_1}{\rho_x} + \frac{a_2}{\rho_y} + \frac{a_3}{\rho_z}$$ (47)

with $a_1 + a_2 + a_3 = 1$.

6. Heterogeneity of Emission

In the study of proteins labeled with fluorescent chromophores, one often finds that the fluorescence decay is not a single exponential. This can be explained by a heterogeneity of sites of the bound chromophores. The decay law is a sum of exponential decays, and one may write

$$S^{(k)} = S_0^{(k)} e^{-t/\tau_k} \tag{48}$$

The anisotropy is given by Eq. (9), with

$$f_k = \frac{S_0^{(k)} e^{-t/\tau_k}}{\sum\limits_k S_0^{(k)} e^{-t/\tau_k}} \tag{49}$$

Then, in the expression of $r(t)$, the fluorescence time constant appears. This expression is simplified and becomes independent of the fluorescence parameters τ_k if the $r_k(t)$ are the same for all the sites.

B. Applications

1. Proteins

These studies have generally been performed with proteins labeled with synthetic chromophores. The first protein investigated was bovine serum albumin labeled with the dansyl fluorescent group [38]. At isoionic pH and at 20°C, the anisotropy is characterized by a single correlation time of value 44 ns, which must be interpreted as the harmonic mean θ_h. At this pH the molecule is compact. At acid pH the anisotropy decay is no longer a single exponential. It can be interpreted with two correlation times of 10 and 60 ns. The first comes from an internal rotation. It is known that at this pH the molecule has a looser structure.

β lactoglobulin A is a globular protein, the basic subunit of which consists of a single polypeptide chain with a molecular weight of 18,400. The single-chain subunits are known to associate with one another to form specific aggregates, the sizes of which are functions of pH and temperature. The dansylated protein has a correlation time of 10 ns at 20°C [61]. In the pH region from 4 to 7.7 the protein exists predominantly as a dimer, and its correlation time is 23 ns. This value is in good agreement with the harmonic mean of the correlation times of an ellipsoid having twice the monomer volume, and with an axial ratio of two.

The correlation times of several proteins labeled with fluorescent dyes have been gathered by Yguerabide et al. [45]. It is found that the measured correlation times are sensibly higher than the correlation times computed for spherical molecules having the same volume. The ratio of these two quantities generally remains about two. Taking account of a possible factor of molecular elongation, it follows that the hydration implied by these results is higher than the hydration obtained from classical hydrodynamic measurements or low-angle X-ray scattering.

Rabbit gamma globulin, IgG, and its enzymic fragments have received special attention. Various experimental results indicate the presence of some flexibility of the hinge region joining the F_{ab} and F_c fragments in the intact molecule. This flexibility is expected to facilitate the formation of the antibody-antigen complex [62,63,64].

The anisotropy decay factor of the dansylated IgG molecules may be characterized by two correlation times; the first has a value of 4 ns and must be attributed to a local motion around the point of fixation of the dye. The value of the second correlation time is 100 ns at 20°C, which corresponds to a macromolecular motion [18,44].

The dansylated F_{ab}, $F_{(ab)_2}$, F_c have the same behavior. The macromolecular correlation times are, respectively, 50, 80, and 60 ns [18]. These values exceed by three times the correlation times of spheres having the same dry volumes. If one takes into account that the F_{ab} and F_c fragments are compact molecules, and if one admits an axial ratio between two and three [62,65], one calculates

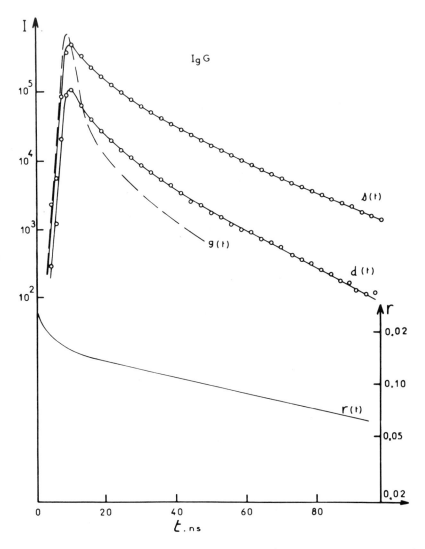

FIG. 2. Dansylated gamma globulin. g(t), exciting flash; s(t) and d(t), fluorescence decay; r(t), anisotropy decay [18].

that the hydration is about 50%. This value is comparable to the hydration of the crystals of the myeloma protein fragments [66,67,68]. On the other hand, the low-angle X-ray scattering technique provides a smaller hydration [65].

It is interesting to compare these data with the results obtained in using the static polarization technique. The isothermal depolarization curves

$$\frac{1}{r} = f\left(\frac{1}{\eta}\right)$$

of the dansylated IgG and its fragments have a curvature which has been attributed to the simultaneous influence of local and macromolecular motions [17,69,70]. From the slope of the isotherm depolarization curve obtained at the high values of T/η, Nezlin et al. [71] determine a correlation time of IgG equal to 20 ns. This value is very small compared to the size of the molecule and is quite similar to the F_{ab} correlation time measured by the same method. The conclusion of the authors is that F_{ab} may rotate freely in the intact molecule.

If one uses the results of the anisotropy measurements, one can show that such a correlation time is actually an average of local and macromolecular correlation times and that the above interpretation is not justified.

Yguerabide et al. [45] have prepared an antibody against the dansyl group. They have measured the anisotropy decay of the complex formed by the fluorescent hapten with the intact antibody and with the F_{ab} fragment.

The F_{ab} fragment has a correlation time of 38 ns at 20°C. The anisotropy decay of the IgG complex has been fitted with two correlation times of 38 and 192 ns. It is concluded that the F_{ab} has a certain freedom of rotation in the intact molecule. One may notice that the 192 ns correlation time has not been directly measured in this work, since the range of time explored is too short. We think that further experiments are needed to ascertain the validity of this analysis.

2. Nucleic Acid-Ethidium Complexes

Much evidence supports the hypothesis of ethidium bromide (EB) intercalation in the double helix of nucleic acids. The dye is highly fluorescent in the complex of EB with DNA. The anisotropy decay of this complex has been measured [72] and the results fitted with two correlation times: one of 24 ns and the other infinite. The 24-ns correlation time is attributed to a local motion which concerns a local deformation of the DNA double helix.

The t-RNA-EB complex has been studied by Tao et al. [73] who found that its correlation time corresponds to a molecule having the shape of a symmetric ellipsoid of axial ratio 2.5 and containing 1.2 g of water per g of dry nucleotide. This result is in good agreement with the values previously obtained by other methods. For the natural fluorescence of the Y base, Beardsley et al. [74] found a much shorter correlation time, which is caused by some local motion in the anticodon site of the molecule.

3. Membranes

The protein phase of the membrane fragments derived from the electric organ of Electrophorus electricus has been studied, when labeled with the dansyl group [75]. After an initial fast decay, the anisotropy remains constant during the time domain explored. This result shows that the proteins are strongly immobilized in the membrane phase. When the membrane is destroyed by a detergent, the protein is solubilized and a strong depolarization occurs. Similar results are obtained when ANS (1-anilinonaphthalene 8-sulfonate) is used as fluorescent probe.

4. Synthetic Polymers

Fluorescent polymers have been obtained by copolymerization of styrene with styryl phenyl anthracene. The proportion of this fluorescent residue has been kept below 2.5% [51,76]. Radical and anionic polymers have been studied in solvents of various viscosities obtained by mixing chlorobenzene with diphenyl chloride (Fig. 3). The time course of the anisotropy may be reproduced by using the formula relative to the symmetric ellipsoid having a transition moment perpen-

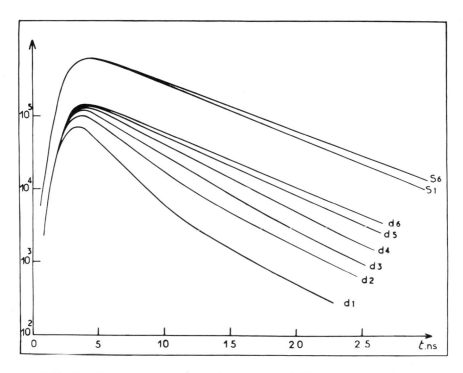

FIG. 3. Fluorescence decay in polarized light of an anionic polystyrene labeled with the styryl diphenyl anthracene residue. The solvents are mixtures of chlorobenzene and pyralene, the viscosities of which vary from 0.8 (curve 1) to 39 cp (curve 6). The decay of the whole fluorescence s(t) shows little variation with the viscosity, while the d(t) curves vary considerably. This is due to the proportionality of the correlation times with solvent viscosity [51].

dicular to the long axis. The diameters of the ellipsoid are 13 and 65 Å long, which corresponds to a polymer segment of 30 monomer residues. The formula contains two correlation times, respectively, characteristic of the motion around and perpendicular to the polymer chain.

V. ENERGY MIGRATION

The origin of concentration depolarization has been attributed by Perrin [33] to energy transfer between like molecules. The cal-

culations have been developed by Förster [34] on the basis of the
theory of transfer by dipolar resonance (for a homogeneous solution
of small molecules). The calculation of depolarization by transfer
already leads to mathematical difficulties. Several approximations
have been proposed [34,35,77,78,79]. These are primarily concerned
with the static anisotropy measured under steady excitation. In the
following section, we are interested in anisotropy decay, and we
will especially consider the case of transfer occurring between
chromophores bound to the same macromolecule.

A.   Theory

1.   General Formula

Here, we will derive formulas which may easily be deduced from
the Förster theory of transfer and the general expression of the
time-dependent anisotropy. We assume that the transfer occurs between molecules of identical species situated in an identical environment. Under those conditions, all the excited molecules have
the same rate constant of emission, and consequently the whole fluorescence decay $S(t)$ is not modified by the transfer processes. $S(t)$
is identical to the decay observed in a very dilute solution [34].
We only consider dipole-dipole energy transfer in the arc of the very
weak interaction [80]. Finally, we assume that the electronic transitions involved in the excitation and emission processes are each
characterized by a single transition moment. The rate of energy
transfers between a donor molecule 1 and an acceptor molecule 2 is
given by the well-known expressions [34]:

$$F_{12} = A \frac{\kappa^2}{R^6} \tag{50}$$

$$A = \frac{9 \log 10}{128\pi^5 n^4 N' \tau_0} \int \frac{F(\tilde{\nu})\epsilon(\tilde{\nu})}{\tilde{\nu}^4} d\tilde{\nu} \tag{51}$$

N' is the Avogadro number for a millimole, n the refractive index,
τ_0 the radiative lifetime, $\epsilon(\tilde{\nu})$ the molar coefficient of absorption,
$F(\tilde{\nu})$ the fluorescence quantum spectrum, and $\tilde{\nu}$ the wave number. The

integral is restricted to the $S_0 \to S_1$ transition. It is called the overlap integral. R is the distance between the chromophores, and κ an angular factor defined by

$$\kappa = \cos \varphi - 3 \cos \varphi_1 \cos \varphi_2$$

where φ is the angle between the transition moments of the acceptor and the donor molecules, and φ_1 and φ_2 are the angles of these transition moments with the direction $\vec{R}$. The transition moments concerned here are those corresponding to the $S_1 \to S_0$ electronic transitions. In a solution, this process of transfer may go on several times between neighboring molecules, leading to an energy migration from a molecule 1 to a molecule n. The process will be stopped if deactivation by radiation or by another means occurs.

Following Knox [35], we will consider the cluster constituted by molecule 1 primarily excited by light absorption and by the set of molecules which can receive this excitation by means of the step-by-step process. A solution must be considered as a mixture of clusters of different species "i." Each species is characterized by the set of coordinates of the excitable molecules in the molecular axes linked to the primarily excited molecule 1. The emission anisotropy is given by Eq. (8)

$$r(t) = f_i r_i(t)$$

where $r_i(t)$ is the emission anisotropy of an isotropic set of clusters "i," and f_i is the fraction of these clusters. The probability that molecule 1 initially excited at time zero is still excited at time t may be written as

$$\rho_1(t) = W_1(t)P(t) \tag{52}$$

where $P(t)$ is the probability that the excitation is still at time t on one molecule of the cluster. $P(t)$ is proportional to the decay $S(t)$. $W_1(t)$ is the conditional probability that the excitation is on molecule 1, knowing that the excitation is on one of the molecules of the cluster. Similarly, the probability that a molecule μ of the cluster is excited at time t is

$$\rho_\mu(t) = W_\mu(t)P(t) \qquad\qquad (53)$$

One has evidently

$$\sum_{\mu=1}^{N} W_\mu(t) = 1$$

$$\sum_{\mu=1}^{N} \rho_\mu(t) = P(t)$$

where the sum concerns all the molecules of a given cluster. $W_\mu(t)$ satisfies the following set of differential equations:

$$\frac{dW_\mu(t)}{dt} = \sum_{\nu \neq \mu} F_{\mu\nu}[W_\nu(t) - W_\mu(t)] \qquad \begin{matrix} \mu = 1, \ldots, N \\ \nu = 1, \ldots, N \end{matrix} \qquad (54)$$

where $F_{\mu\nu} = F_{\nu\mu}$ is the transfer rate constant between the molecules μ and ν. From (50), one gets

$$F_{\mu\nu} = A \frac{\kappa_{\mu\nu}^2}{R_{\mu\nu}^6} \qquad\qquad (55)$$

The set of equations (54) is easily deduced from the set given by Förster [34] which the $\rho_\mu(t)$ obeys. The emission anisotropy of the cluster of species i is

$$r_i = \sum_{\mu=1}^{N} \frac{3 \cos^2 \Phi_{\mu 1}^{(i)} - 1}{5} W_\mu^{(i)}(t) \qquad\qquad (56)$$

where $\Phi_{\mu 1}^{(i)}$ is the angle between the absorption transition moment of molecule 1 and the emission transition moment of molecule μ. In an isotropic solution of small molecules, the probability density function giving the emission transition moments $\vec{E}_\mu$ ($\mu \neq 1$) around the emission moment $\vec{E}_1$ is independent of the azimuthal angle of $\vec{E}_\mu$ around $\vec{E}_1$. It can easily be shown that this property allows the emission anisotropy to be written

$$r(t) = r_0 \sum_i \sum_\mu \frac{3 \cos^2 \theta_\mu^{(i)} - 1}{2} W_\mu^{(i)}(t) \qquad\qquad (57)$$

where r_0 is the fundamental anisotropy and θ_μ the angle between $\vec{E}_1$ and $\vec{E}_\mu$.

2. Chromophores Bound to Macromolecules

If the solution of macromolecules is dilute, energy migration only occurs between chromophores bound to the same macromolecule. There are two specific cases, namely: the presence of a possible preferential orientation and the existence of the Brownian motion.

a. __Preferential Orientation__. In this case, Eq. (57) is no longer valid. Let us assume, for instance, that the chromophores are distributed along a regular helix. In addition, we assume that there is no Brownian motion. One may show that (see Appendix)

$$r(t) = (r_0 - 0.1) \sum_\mu (2 \cos^2 \theta_\mu - 1) W_\mu(t) + 0.1 \tag{58}$$

Equation (58) predicts that for $t \to \infty$, $r(t)$ is equal to 0.1, while under the same conditions $r(t) = 0$ with Eq. (57).

b. __Brownian Motion__. We will distinguish between the Brownian motion of the whole macromolecule and the local Brownian motion. Weber and Anderson [81] have studied the influence of excitation transfers on the average correlation times measured by the static method. In their derivations these authors do not consider the time dependence of the transfer phenomenon. Fayet and Wahl [31] have shown that the presence of transfers accelerates the anisotropy decay and leads to an apparent correlation time smaller than the actual one. They derived an expression for the transfer contribution to the anisotropy decay, which is based on Jablonski's approximation.

The Brownian motion of a big macromolecule may be neglected, but a local Brownian motion may occur around an average position of the chromophore. If the chromophore is symmetrically distributed around its average position (that is to say, if the distribution is independent of the azimuthal angle), then the emission anisotropy may be written as

$$r(t) = r_T(t) r_B(t) \tag{59}$$

where $r_B(t)$ is the Brownian anisotropy measured in the absence of transfers and r_T the anisotropy of transfer in the absence of the Brownian motion divided by r_0:

$$r_T(t) = \frac{1}{r_0} \Sigma r_i f_i$$

with the following approximate expression:

$$r_i(t) = r_0 \Sigma \frac{3 \cos^2 \delta_\mu^{(i)} - 1}{2} W_\mu^{(i)}(t) \tag{60}$$

where δ_μ is the angle between the average position of the molecules initially excited and the molecule secondarily excited μ of a given cluster i. $W_\mu^{(i)}(t)$ is also calculated for the average angular position.

B. Application to the
Study of the Complex DNA-EB

It has been shown that the slope of the anisotropy decay increases with the ratio D/P of the number of dye molecules per nucleotide. A quantitative interpretation has been obtained of the anisotropy decay by using an approximation of Förster's [32]. In this calculation, the chromophore distribution is assumed to be random, continuous, and linear along the DNA molecule. Energy may only be transferred back and forth between the primarily excited molecule and its nearest neighbor. With this approximation and for values of D/P relatively small, the experiments are satisfactorily described. For high values of D/P, a Monte Carlo calculation is used [36,36a]. According to Paoletti and Le Pecq [82], one generates a statistical distribution of chromophores among equivalent sites, with the condition that two adjacent sites cannot be simultaneously occupied, that is the excluded site model. One assumes that the intercalation of a dye molecule brings about an elongation of 3.4 Å

and an angular deformation characterized by the angle δ. One speaks
of winding or unwinding, accordingly as δ is positive or negative.

The time course of the energy migration is simulated for a set
of values of δ. The average value of the anisotropy $r_T(t)$ is then
computed for a great number of simulations. $r(t)$ is computed by
using Eq. (59), where $r_B(t)$ is the decay measured for a very small
value of D/P.

When comparing the experimental curves with the computed ones,
one finds that the best fit is obtained for δ = -16 ± 4 deg. This
result is in agreement with the works of Fuller and Waring [83] and
of Bauer and Vinograd [84] who found that ethidium bromide induces
an unwinding of the DNA helix.

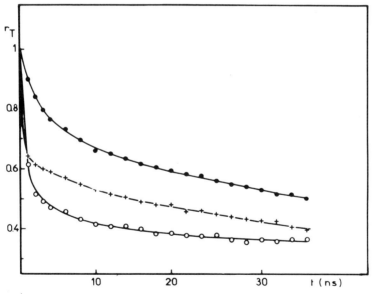

FIG. 4. Curves computed by the Monte Carlo method representing
$r_T(t)$, the energy migration contribution to the anisotropy decay of
the EB-DNA complex with D/P = 0.136. The curves differ by the value
assumed for the deformation angle δ induced by the insertion of an
EB molecule in the DNA helix. (...), δ = -20 deg; (+++), δ = -24 deg;
(ooo), δ = -6 deg. (Genest and Wahl, unpublished results.)

On the contrary, Paoletti and Le Pecq [82] have proposed a
winding angle of 14 deg based on their static polarization measurements. One may notice that static polarization measurements are a
less accurate means of determining δ than anisotropy decay measurements. An important change in the shape of the anisotropy decay
curve may lead to a small change in the static anisotropy curve because a compensation occurs when one performs the integration expressed by Eqs. (16) and (17).

The experiments quoted above were carried out with calf thymus
DNA. For poly d(A-T)-EB complexes isolated from <u>Cancer</u> <u>pagurus</u>, the
unwinding angle was found to be -17 ± 2 deg, a value practically
equal to the one of calf thymus DNA [85].

The method was also applied to poly(rA-rU)-EB complexes [86].
This polyribonucleotide has an A helix structure in contrast with
the DNA B structure. That specific A structure had to be taken into
account in order to obtain a calculated anisotropy curve which fitted
the experimental data. The unwinding angle which was found was equal
to -38 deg.

This result is in agreement with the model studies which predicted that an EB molecule induces a larger unwinding angle in the
A structure than in the B structure [87]. In addition, the results
provide information about the angle between the transition moment of
the intercalated dye and the line joining the pyrimidine-purine
centers.

VI.   CONCLUSION

It has become apparent that the fluorescence anisotropy decay
measurements of a chromophore bound to a macromolecule or to a biological system allow a more correct and accurate analysis of the
molecular phenomena than the determination of the static polarization.

On the one hand, by the study of the Brownian depolarization, one
obtains information on the flexibility and the size of the macromolecule, as well as on the local motions of the chromophores. On
the other hand, the study of the depolarization by transfers provides accurate data on the spatial disposition of a set of identical
chromophores. These measurements become possible, thanks to the
development of the single photoelectron counting method.

Generally speaking, this method allows and will allow new information to be obtained in various systems of biological interest.
One may predict that the number of those who will use this method
will increase considerably in the next few years.

APPENDIX: ENERGY MIGRATION IN AN ARRAY OF
CHROMOPHORES DISTRIBUTED ALONG A REGULAR HELIX

We assume that the helix length is infinite. We designate the
chromophores which follow the primarily excited molecule 1 by the
indices 2, 3, ... and the chromophores which precede the molecule 1
by the indices -2, -3, ...

One has

$$\Phi_\mu = \lambda + \theta_\mu$$

λ is the angle between the absorption and the emission moments of
the molecule 1 and θ_μ is the angle between the emission moments of
molecule 1 and μ. If $\mu' = -\mu$, one has evidently

$$\Phi_{\mu'} = \lambda - \theta_\mu$$

and

$$W_\mu(t) = W_{\mu'}(t)$$

Equation (56) of the anisotropy can be written

$$r(t) = \sum_{\mu=2}^{\infty} \frac{3(\cos^2 \Phi_{\mu} + \cos^2 \Phi_{\mu'}) - 2}{5} + r_0 W_1(t) \qquad (61)$$

An elementary calculation gives

$$\cos^2 \Phi_{\mu} + \cos^2 \Phi_{\mu'} = (2 \cos^2 \lambda - 1)(2 \cos^2 \theta_{\mu} - 1) + 1 \qquad (62)$$

Then Eq. (61) becomes

$$r(t) = \sum_{\mu=2}^{\infty} \frac{3(2 \cos^2 \lambda - 1)(2 \cos^2 \theta_{\mu} - 1) + 1}{5} W_{\mu}(t)$$
$$+ r_0 W_1(t) \qquad (63)$$

which can also be written as

$$r(t) = \sum_{-\infty}^{+\infty} \frac{3(2 \cos^2 \lambda - 1)(2 \cos^2 \theta_{\mu} - 1) + 1}{10} W_{\mu}(t) \qquad (64)$$

Finally, taking into account the relation (23) between r_0 and $\cos^2 \lambda$, one obtains Eq. (58).

REFERENCES

1. F. Weigert, Verhandel. Deut. Ges. Phys., 23, 100 (1967).

2. F. Perrin, J. Phys., 7, 390 (1926).

3. F. Perrin, Ann. Phys. (Paris), 12, 169 (1929).

4. F. Perrin, J. Phys. (Paris), 5, 497 (1934).

5. F. Perrin, J. Phys. (Paris), 7, 1 (1936).

6. F. Perrin, Acta Phys. Polon., 5, 335 (1936).

7. A. Jablonski, Bull. Acad. Pol. Sci. Ser. Sci. Math. Astr. Phys., 8, 259 (1960).

8. N. S. Marinesco, J. Chim. Phys., 24, 593 (1927).

9. E. Gaviola, Z. Physik., 42, 853 (1927).

10.   P. Pringsheim and H. Vogels, J. Chim. Phys., 33, 261 (1936).

11.   R. K. Bauer and T. Szczurek, Acta Phys. Polon., 22, 29 (1962).

12.   G. Weber, Biochem. J., 51, 145, 165 (1952).

13.   G. Weber, Adv. Protein Chem., 8, 415 (1953).

14.   R. F. Chen, H. Edelhoch, and R. F. Steiner, in Physical Prin-
      ciples and Techniques of Protein Chemistry (S. I. Leach, ed.),
      Part A, Academic Press, New York, 1969, p. 171.

15.   W. Dandliker and A. J. Portmann, in Excited States of Protein
      and Nucleic Acids (R. F. Steiner and I. Weinryb, eds.), Mac-
      millan, New York, 1971, p. 199.

16.   Ph. Wahl, Thesis, Strasbourg, 1962.

17.   Ph. Wahl and G. Weber, J. Mol. Biol., 30, 371 (1967).

18.   J. C. Brochon and Ph. Wahl, Eur. J. Biochem., 25, 20 (1972).

19.   A. Jablonski, Z. Phys., 95, 53 (1935).

20.   A. Jablonski, Z. Phys., 106, 526 (1936).

21.   A. Jablonski, Z. Naturforsch., 16a, 1 (1961).

22.   Szymanowski, Z. Physik, 95, 466 (1935).

23.   W. Kessel, Z. Physik, 103, 125 (1936).

24.   R. K. Bauer, Z. Naturforsch., 18a, 718 (1963).

25.   R. D. Spencer and G. Weber, J. Chem. Phys., 52, 1654
      (1970).

26.   J. B. Birks and I. H. Munro, Progr. Reaction Kinetics, 4, 231
      (1967).

27.   W. R. Ware, in Creation and Detection of the Excited States
      (A. A. Lamola, ed.), M. Dekker, New York, 1971, p. 213.

28.   Ph. Wahl, in New Techniques in Biophysics and Cell Biology
      (R. Pain and B. Smith, eds.), Wiley, UK, 1975.

29.   I. Isenberg, this book.

29a. J. Y. Yguerabide, in Methods in Enzymology, Vol. 26, Part C
      (Hirs and Timasheff, eds.), Academic Press, New York, 1972,
      p. 98.

30.   Ph. Wahl, in Protides of Biological Fluids, Vol. 19 (Peeters, ed.), Pergamon Press, England, 1972, p. 393.

30a. R. Rigler and M. Ehrenberg, Quarterly Rev. Biophys., 6, 139-199 (1973).

31.   M. Fayet and Ph. Wahl, Biochim. Biophys. Acta, 181, 373 (1969).

32.   D. Genest and Ph. Wahl, Biochim. Biophys. Acta, 259, 175 (1972).

33.   J. Perrin, 2eme Conseil de Chimie Solvay, Gauthier Villars, Paris, 1925, p. 322.

34.   Th. Förster, Ann. Physik, 2, 55 (1948).

35.   R. S. Knox, Physica, 39, 361 (1968).

36.   D. Genest and Ph. Wahl, in Dynamical Aspects of Conformation Changes in Biological Macromolecules (C. Sadron, ed.), D. Reidel, The Netherlands, 1973.

36a. D. Genest and Ph. Wahl, Biophys. Chem., 1, 266-278 (1974).

37.   Ph. Wahl, C. R. Acad. Sci., 260, 6891 (1965).

38.   Ph. Wahl, C. R. Acad. Sci., 263, 1525 (1966).

38a. Ph. Wahl, J. C. Auchet, and B. Donzel, Rev. Sci. Instrum., 45, 28-32 (1974).

39.   I. Isenberg and R. Dyson, Biophys. J., 9, 1337 (1969).

40.   W. Knight and B. A. Selinger, Spectrochim. Acta, 27A, 1223 (1971).

40a. B. Valeur and J. Moirez, J. Chim. Phys., 70, 500-506 (1973).

40b. A. Grinwald and I. Steinberg, Anal. Biochem., 59, 583-598 (1974).

41.   Y. Koechlin, Thesis, Paris, 1961.

42.   L. M. Bollinger and G. E. Thomas, Rev. Sci. Instr., 32, 1044 (1961).

43.   T. Tao, Biopolymer, 8, 609 (1969).

44.   Ph. Wahl, Biochim. Biophys. Acta, 175, 55 (1969).

45.   P. Yguerabide, H. F. Epstein, and L. Stryer, J. Mol. Biol., 51, 573 (1970).

46.   R. S. Schuyler and I. Isenberg, Rev. Sci. Instr., 42, 813
      (1971).

47.   G. Weber, J. Chim. Phys., 55, 878 (1958).

48.   A. C. Albrecht, J. Mol. Spect., 6, 84 (1961).

49.   F. Dorr, in Creation and Detection of the Excited States (A.
      A. Lamola, ed.), Vol. I, Part A, Chap. 2, M. Dekker, New York,
      1971, p. 53.

50.   P. Soleillet, Ann. Phys. (Paris), 12, 23 (1929).

51.   Ph. Wahl, G. Meyer, and J. Parrod, Eur. Polymer J., 6, 585
      (1970).

52.   I. Z. Steinberg, this book.

53.   A. Jablonski, Acta Phys. Polon., 10, 193 (1950).

53a. Ph. Wahl, Chem. Phys., 7, 210 (1975); 7, 220 (1975).

54.   S. C. Harvey and H. C. Cheung, Proc. Natl. Acad. Sci. U.S.A.,
      69, 3670 (1972).

55.   G. G. Beldford, R. L. Beldford, and G. Weber, Proc. Natl. Acad.
      Sci. U.S.A., 69, 1392 (1972).

55a. T. J. Chuang and K. B. Eisenthal, J. Chem. Phys., 57, 5094
      (1972).

55b. M. Ehrenberg and R. Rigler, Chem. Phys. Lett., 14, 539 (1972).

56.   R. Memming, Z. Phys. Chem., 28, 168 (1961).

57.   Y. Gottlieb and Ph. Wahl, J. Chim. Phys., 60, 849 (1963).

58.   D. Wallach, J. Chem. Phys., 47, 5258 (1967).

59.   E. Dubois Violette, F. Geny, L. Monnerie, and O. Parodi, J.
      Chim. Phys., 66, 1865 (1969).

60.   R. F. Steiner and McAlister, J. Polymer Sci., 24, 107 (1957).

61.   Ph. Wahl and S. N. Timasheff, Biochemistry, 8, 2945 (1969).

62.   M. E. Noelken, C. A. Nelson, C. E. Buckley, and C. Tanford,
      J. Biol. Chem., 240, 218 (1965).

63.   A. Feinstein and A. J. Rowe, Nature, 205, 147 (1965).

64. R. C. Valentine and N. M. Green, _J. Mol. Biol._, 27, 615 (1967).

65. I. Pilz, G. Puchschwein, O. Kratky, M. Herbst, O. Haager, W. E. Gall, and G. M. Edelman, _Biochem._, 9, 211 (1970).

66. R. L. Humphrey, _J. Mol. Biol._, 29, 525 (1967).

67. D. J. Goldstein, R. L. Humphrey, and R. J. Poljak, _J. Mol. Biol._, 35, 247 (1968).

68. R. J. Poljak, H. M. Dintzis, and D. J. Goldstein, _J. Mol. Biol._, 24, 351 (1967).

69. J. A. Weltman and G. M. Edelman, _Biochemistry_, 6, 1437 (1967).

70. Y. A. Zagyanski, R. S. Nezlin, and L. A. Tumerman, _Immuno Chem._, 6, 787 (1969).

71. R. S. Nezlin, Y. A. Zagyanski, and L. A. Tumerman, _J. Mol. Biol._, 50, 569 (1970).

72. Ph. Wahl, J. Paoletti, and J. B. Le Pecq, _Proc. Natl. Acad. Sci._, 65, 417 (1970).

73. T. Tao, J. H. Nelson, and C. R. Cantor, _Biochemistry_, 9, 3514 (1970).

74. K. Beardsley, T. Tao, and C. Cantor, _Biochemistry_, 9, 3524 (1970).

75. Ph. Wahl, M. Kasai, and J. P. Changeux, _Eur. J. Biochem._, 18, 332 (1971).

76. Ph. Wahl, G. Meyer, and J. Parrod, _C.R. Acad. Sci. Série C_, 264 1641 (1967).

77. G. Weber, in _Fluorescence and Phosphorescence Analysis_ (D. Mercules, ed.), Interscience, New York, 1966, p. 217.

78. A. Jablonski, _Acta Phys. Polon._, 14, 295 (1955).

79. A. Jablonski, _Acta Phys. Polon._, 17, 481 (1958).

80. Th. Förster, in _Modern Quantum Chemistry III_ (O. Sinanoglu, ed.), Academic Press, New York, 1965, p. 93.

81. G. Weber and S. Anderson, _Biochemistry_, 8, 361(1969).

82. J. Paoletti and J. B. Le Pecq, _J. Mol. Biol._, 59, 43 (1971).

83. W. Fuller and M. J. Waring, _Ber. Bansenges Phys. Chem._, 68, 805 (1964).

84. W. Bauer and J. Vinograd, <u>J. Mol. Biol</u>., <u>33</u>, 141 (1968).

85. J. L. Tichadou, D. Genest, Ph. Wahl, and G. Aubel Sadron, <u>Bio-</u><u>phys. Chem</u>., in press, 1975.

86. Ph. Wahl, J. L. Tichadou, and D. Genest, in preparation.

87. W. J. Pigram, W. Füller, and L. D. Hamilton, <u>Nature (N. Biol.)</u>, <u>235</u>, 17 (1972).

Chapter 2

TIME DECAY FLUOROMETRY BY PHOTON COUNTING

Irvin Isenberg
Department of Biochemistry and Biophysics
Oregon State University
Corvallis, Oregon

I. INTRODUCTION

Three laboratories simultaneously and independently introduced
monophoton counting to measure fluorescent time decay [1,2,3]. The
interesting thing is not so much that all of these laboratories developed a procedure simultaneously; it often happens that different
laboratories develop the same thing at the same time. What is interesting is that these three groups introduced photon counting for
three quite different purposes. Koechlin [3] needed a way of measuring short lifetimes. Bennett [1] had low intensities of emission,

and Bollinger and Thomas [2] wanted a means of studying a wide range
of decay. That all three could use the same basic procedure says
a great deal about that procedure. It is now clear that there are
even further advantages to monophoton instruments than what were
known in 1961; these will become evident later in this chapter.

With all of these advantages, one might imagine that, at last,
an ideal instrument is at hand. Unfortunately, this is not the
case. The capabilities of an instrument must be judged in the light
of what we would like it to do. And, at least for a biochemist or
biophysicist, the capabilities of the instrument do not yet come up
to the ideal requirements.

What is the problem? The difficulty is that, like most data,
fluorescence time decay data are not, of themselves, interesting.
What is of interest is the decay parameters which are not themselves
pieces of data but are quantities which are derived from the data.
The typical biochemical problem that we would like to solve is one
involving many parameters of decay; the one that we are able to
solve, at least right now, has only a few parameters.

The data analysis problem is one that is characterized as "ill-
conditioned." The solution of an ill-conditioned problem is, by
definition, very sensitive to errors in the data. This implies that
the data themselves must be very precise, and the instrument to take
the data must have few errors. Linked to the instrument is the
method of data analysis which must not, of itself, introduce error
or bias the analysis.

There is a further implication. In any problem it is necessary,
of course, to use some type of criterion for having confidence in
the decay parameters that are obtained. For an ill-conditioned
problem it is important to use such criteria rigorously and explicitly. These criteria may vary from one problem to another, or
even from one laboratory to another, but some set of meaningful
definite criteria must be used.

As will be seen, the criterion of simple visual observation cannot be used. Yet this is the first test that workers are accustomed to use, not only because it is a simple test but because, one hastens to add, it usually works. The test, of course, is to draw figures with both data points and theoretical curves using computed parameters. If the data points cluster nicely about the theoretical curve, the parameters are then acceptable.

Such a procedure will simply not work with time-decay problems. Figure 1 shows why. In this figure a computer-simulated fluorescence decay is shown. The decay therefore contains known parameters and the "instrument" is not only good, it is ideal in that the only error is a counting error (see below). Also shown is a smooth curve which was drawn with incorrect parameters. To the eye the fit looks good. The parameters, however, are bad.

The eye is simply not a good judge of the fit of time-decay curves. Fortunately, it is possible to present criteria, and later in this chapter a number of these, which we have found useful, will be presented.

Because the data analysis problem is ill-conditioned, the instrument that collects the data must be good. The goals of the technique, the instrumentation, and the analysis of data are all interwoven. There have been many flash fluorometers developed, but presently monophoton counting appears to be the method of choice. It has, both in principle and in practice, distinct advantages over other existing flash techniques [4].

In this chapter we shall consider only the monophoton technique. However, it must be noted that the modulation method [5] is also capable of great precision. This technique measures the response to modulated exciting light rather than the response to a flash. It is likely that both methods will find their appropriate niches in biochemical studies, but it is too early to attempt to give a description of what these niches may be.

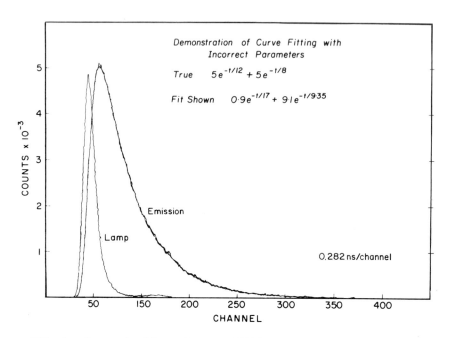

FIG. 1.   Demonstration of curve fitting with incorrect parameters. A convolution was taken of the response to a δ function, f = 5 exp (-t/12) + 5 exp(-t/8) and the excitation. Noise was added by means of a random number generator. The resultant curve is shown as the noisy curve marked emission. The smooth curve through the emission data is a convolution of the excitation and f = 0.9 exp(-t/17) + 9.1 exp(-t/9.35).

The recent excellent and thorough review by Ware [4] permits us to be highly selective in this chapter. No attempt will be made to give a survey, or history, of flash or even monophoton instruments. The chapter will present a general description of monophoton counting, but it will emphasize those features that our own laboratory has found to be useful, both instrumentally and computationally. Thus the reader should be cautioned that the present chapter is neither well-rounded nor exhaustive, but contains a distinct bias.

II. INSTRUMENTATION

A. Principle of the Method

In the monophoton method a fluorescent sample is excited by a short pulse of light. The sample, in turn, emits many photons. One of these photons is randomly selected by the instrument, and its emission time is recorded. The process is then repeated many times. Eventually enough events are recorded to yield an estimate of the emission curve. |

The instrument must have the following features:

1. A stable exciting lamp. The term "stable" means that, ideally, every pulse must have the same shape as every other pulse.

2. A clock to measure the time of emission and a means of storing the data.

3. A means of detecting single photons and of guaranteeing that only single, and not multiple, photon events are recorded.

4. Low noise.

B. The Lamp

There have been a variety of flash lamps developed dating back to that of Malmberg [6], who was apparently the first to produce nanosecond lamps. His lamp used a spark in hydrogen at a pressure of 100 mm Hg. Others have studied similar lamps having a variety of gases [7,8,9]. High-pressure hydrogen lamps with mercury-wetted electrodes have also been used [10,11,12]. Innes and Kerns [13] developed a lamp having a spark between a tungsten wire and barium titanate. Commercial coaxial line has also been used by Tao [14].

Our laboratory has tried a number of these lamps. We have been
most successful with the lamp described by Ware [4,9]. Ware's lamp
is simple to construct and use; it provides a narrow pulse, and it
is stable for long periods of time.

The lamp uses platinum electrodes. The negative electrode is
a ball, about 2 mm in diameter, formed in a gas-oxygen flame; the
positive electrode is a sharp point formed by grinding and polishing.
We generally use a gap width of 1-2 mm. Wide gaps give more light,
but require a higher operating voltage which, in turn, may lead to
severe radiofrequency noise problems, as discussed below.

Lamps may be run in two modes: free running or pulsed. In the
free-running mode, one simply applies a high voltage to the electrodes through a high resistor and across a capacitor. The capacitor may be, in fact, simply the stray capacitance of the circuit.
The voltage across the electrodes rises until the breakdown field
strength of the gas is reached. After sparking, the process is repeated. The repetition rate is determined by the voltage applied
and the RC constant of the circuit.

Free running has the advantage of simplicity. In the author's
opinion, however, the simplicity of the free-running mode is obtained at a high cost. His reasons are essentially the same as
those given by Ware [4], and, indeed, the author takes this oppor-
tunity to thank Dr. Ware for first educating him on the advantages
of using a pulsed rather than a free-running lamp.

In a free-running lamp, once the lamp is built and filled with
gas, there is very little flexibility of operation left. In a
pulsed system, the lamp pressure, rate of pulsing, and the voltage
applied are all independently adjustable. The ease and flexibility
of operation more than compensate for the extra trouble of building
a pulsing circuit which, in fact, is not difficult to build at all.
In our work we have used the pulser described by Ware [4]. We generally pulse at about 1.5×10^4 Hz and 4 kV.

It should be added that it seems probable that, eventually, laser pulses will be used as an excitation source. Thus one may expect, perhaps before not too long, that discussions of spark gap lamps, and associated matters, will become obsolete.

C.   Measurement of Time Intervals

The clock of a monophoton fluorometer is a time-to-amplitude converter, abbreviated as TAC. A TAC has a start input, a stop input, and an output. When a start pulse is applied, a voltage in the TAC rises linearly with time until a stop pulse is applied. The TAC then puts out a pulse whose magnitude is proportional to the time between the start and stop signals. The TAC is thus the basic timing mechanism of the system.

A commercial TAC will have various time ranges which can be set according to the needs of the particular experiment. A time interval less than the range set may be measured. If, however, no stop signal appears before the end of the range is reached, no output will be produced. The TAC will simply be reset.

Suppose, however, that two (or more) stop pulses are present within the range of the TAC. Only the first pulse will stop the TAC; the second will have no effect. Suppose, further, that the overall frequency of pulsing is so high that, most of the time, two or more pulses occur within the time span of the TAC. The TAC will detect only the first pulse, and it is clear that random sampling will no longer occur. The decay curve will be biased toward shorter times.

Methods of preventing such a bias are obviously needed and are, in fact, essential. These will be discussed subsequently. Meantime, let us assume that the bias is indeed prevented by some method, and let us consider the storage of information.

The data are stored in a (multichannel) pulse-height analyzer.
In this instrument, the time span is divided into channels, and,
ideally, each channel, no matter where it is located, has the same
width in time. Every event detected by the TAC is stored in a particular channel of the pulse-height analyzer. Eventually, with
enough repetition, the pulse-height analyzer will contain a histo-
gram of time-decay data. Output of the data can then be effected
in any of a number of ways for subsequent data analysis.

D. Photomultiplier and Pulse Timing

Photomultiplier tubes that are useful for monophoton counting
must have a number of stringent requirements. They must be high-
gain tubes so that single photons may be detected successfully. In
addition, however, the thermal noise must be low, because electrons
emitted thermally can produce the same type of pulse as electrons
produced by photons.

The use of energy windowing [15] will be described later in
this chapter. It should be noted here, however, that for such a
use, a high photoelectron pulse-height resolution is needed. This
means that, ideally, single photons should produce one photocurrent,
and double photons should produce a higher photocurrent, well separated from the monophoton photocurrent. Actually there will, of
course, be a range of currents for single and double photons. For
use in energy windowing, however, the ranges should not overlap much.
We have found the RCA "quanticon" tubes to be useful for energy windowing, and we routinely use RCA 8850 photomultipliers.

To obtain good pulse-height resolution, it is important to
critically adjust the focusing between the cathode and first dynode
[16]. It is our experience that this tuning must be done carefully.

It is of some importance to realize that the range of different
photocurrent magnitudes, even for monophotons, will be large. The

pulses will have different sizes, creating a timing problem. If the
time of a pulse were taken to be the time that a pulse reached a
certain voltage, then it is clear that this time would depend on the
magnitude of the pulse. Considerable time jitter would appear,
therefore, in a measurement. More accurate timing is achieved by
the so-called constant fraction of pulse-height trigger [17,18].
In this system the pulse is split into two. One of these pulses is
then inverted and delayed, while the other is attenuated.

Let A be the amplitude of one of the pulses. Then we may represent the two pulses by

$$V_1 = -A\varphi(t - \tau)$$

$$V_2 = kA\varphi(t)$$

where τ is the delay time and k is the attenuation factor. The
time at which

$$V = V_1 + V_2 = 0$$

is independent of the amplitude of the pulse. Therefore, by measuring the time at the zero point of V, constant fraction timing
gets around the problem of having the timing depend on the amplitude
of the pulse.

In practice, careful tuning can permit the timing error to be
kept at the level of tens of picoseconds, when the amplitude of the
pulse changes about a factor of 2.

The timing of an event involves the measurement of two times
because we are interested in an interval of time. The TAC must
have a zero point. The measurement of time zero is not as difficult as the time of receipt of a photon. If a photomultiplier is
used, it can look head-on at the lamp flash and detect the excitation pulse which consists of a large number of photons [15]. Alternatively, a voltage pulse may be picked off from the lamp circuit [4] and used to mark time zero.

E. Single Photons or
Multiple Photons; Energy Windowing

Let us first consider a bad monophoton experiment. Let us suppose that we measure times from t = 0 to t = T. We shall call T the
measuring time span. Let us further suppose that the resolution of
our instrument is so good that every photon impinging on the photomultiplier leads to a separate event, but if two photons are detected
within T, only the first one will activate the TAC. Let us now suppose that most of the time there will be two or more photons falling
within the measured time span. It is clear that the instrument is
not detecting a random sample of the emitted photons. The measured
emission will be biased toward short times. The basic principle of
the measurement has been violated.

The detection of multiple photon events has become known as the
pile-up problem. There are a number of ways of trying to solve the
pile-up problem: one may attenuate the emission to very low levels,
one may attempt to calculate a pile-up correction, or one may try to
avoid measuring any multiple photon events at all.

The first method uses a very low emitted light flux. One simply
cuts the emission to the photomultiplier to such a low level that the
probability of receiving even a single photon during the entire measuring time span is very small. Then the probability of a multiple
event will be even lower. As the intensity is lowered, at some light
level, the number of multiple photon events becomes negligible.

This attenuation procedure has the great advantage of simplicity. Unfortunately, it has severe limitations. Suppose, for con-
creteness, that 10^7 counts are needed for a given experiment, a
requirement that is not unreasonable [19]. Let us further suppose
that the exciting light flashes at a rate of 10^4 pulses per second.
Finally, let us suppose that the emission is attenuated sufficiently
so that only 10^2 events per second are recorded. The experiment
will take over a day to do. During this time the lamp may change
its characteristics, the electronics may drift, and the sample may

change its properties. Furthermore, it is possible that even a 1% occurrence of multiple photon events might not be low enough. We shall call the number of events recorded per exciting lamp flash the efficiency of the system. Even at 1%, the precision of the data might still not be good enough for the problem at hand. One might have to reduce the efficiency and increase the time for an experiment. One cannot do this indefinitely, however. Even if the lamp, electronics, and sample were all stable, as the efficiency was lowered one would reach a point where noise pulses would begin to compete with the signal pulses. Thus, even in principle, one cannot continue to make the data more and more precise by lowering the efficiency.

An attempt at calculating a correction to pile-up has been made by Coates [20]. The principle of the procedure is straightforward. Coates assumes that the events are governed by Poisson statistics. He is then able to calculate a correction. In principle, then, one makes measurements and then corrects the measurements for pile-up.

For the measurement of decay times that are not too short, a computed pile-up correction should certainly permit the use of a higher collection efficiency than what would otherwise be possible. The limitation arises because of the requirement, implicit in the calculation, that all pulses be discrete and resolved by the instrument.

The anode pulses of photomultipliers are a number of nanoseconds wide. If two photons arrive too close together, a single distorted pulse will result. The equations for the correction will no longer apply for such an event. A more refined calculation, taking this into account, could conceivably be made, but, until then, computed corrections appear feasible only for relatively long decays.

Davis and King [21] remove double pulse events by hardware. This is a better procedure than a computed correction because no assumption need be made as to the nature of the statistics. A pulse-pair discriminator sends an inhibit pulse to the TAC whenever two pulses are detected within the measuring time span. However, this technique will also fail for decay times that are too short since, here again, two pulses will condense to a single distorted pulse. However, for long times, the method permits a high collection efficiency.

Our laboratory approached the problem of discriminating against multiple photon events in a direct manner [15]. We measure the energy of the event, and only if the energy is a single photon energy do we record it. Such a procedure may be called energy windowing.

The possibility of making a discrimination between single and multiple photon events [15] was opened by the development of the high-gain dynode [22], now used as the first dynode in the RCA "quanticon" photomultipliers. These tubes have a good pulse-height resolution, so that single and double photon events cover different photoelectron ranges (Fig. 2).

In this system two signals are taken from the photomultiplier, the timing signal from the anode, and a measuring, or linear, signal from one of the dynodes. We have found it convenient to take the linear signal from the 11th dynode of an RCA 8850 tube. This

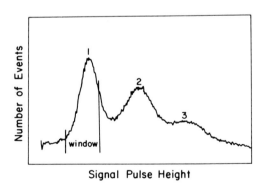

FIG. 2. Pulse-height resolution of an RCA 8850 photomultiplier. A low photon flux falls on the cathode. Number of events as a function of the signal pulse height from the 11th dynode. Peaks due to 1, 2, or 3 photons are shown. Also shown is a typical window used in energy windowing.

linear signal has an amplitude that is proportional to the photocurrent.

It is important that the preamplifier-amplifier combination in
the linear circuit have a time constant that is long compared to
the measuring time span. The time constant in our instrument is
1.25 μsec, while the measuring time span is usually at least an
order of magnitude smaller. Therefore, pulses which are separated
by less than the measuring time will be integrated into a large
pulse. The photomultiplier pulse-height resolution is good enough
so that double events can be detected. It is then not difficult to
arrange the electronics so that such events are not recorded.

Figure 3 shows such an arrangement. The EF (energy fluorescence) channel will send a pulse to the coincidence module only if
the pulse is within the window set by the upper and lower level
controls of the single-channel analyzer.

The single-channel analyzer in the EL (energy, lamp) circuit
has a different function. The window discriminates against low
amplitude noise. This analyzer provides a second signal for the
coincidence module, which results in an enabling pulse that opens
the linear gate whenever a monophoton event and a lamp flash occur
within a reasonable time of one another. This time can be adjusted
in the coincidence module and is chosen to be somewhat larger than
the measuring time span. With this arrangement, the gate will be
opened only to monophoton events.

With the arrangement shown in Fig. 3, therefore, photocurrent
discrimination selects only monophoton events. This permits a
rapid collection of data. Our laboratory customarily uses a lamp
repetition rate of 15,000 pulses per second. We can usually collect
data at about 4000 counts per second. In only 40 min, 10^7 counts
can be obtained, and the collection has still discriminated sharply
against multiple photon events.

At this point we can describe the events that follow a flash
of light (Fig. 3). The 8575 photomultiplier looks at the exciting

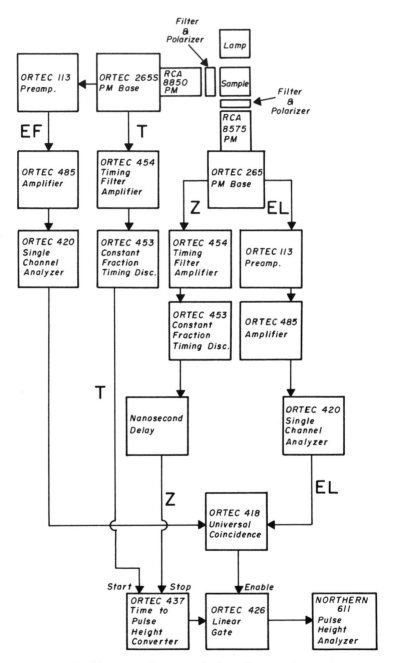

FIG. 3.   Block diagram of a monophoton fluorometer using energy
windowing. Z designates the zero-time channel. EL is the lamp
energy channel. T is the timing channel, and EF the fluorescence
energy channel.

lamp through the sample. At the flash, a pulse travels along the
EL path. If a single photon event occurs (more precisely, if an
event occurs within the bounds set by the single-channel analyzer),
then a coincidence will be registered, and the gate to the pulse-
height analyzer will be opened.

The function of the nanosecond delay in the Z (zero time) circuit is to delay the pulse enough so that the TAC starts with a
fluorescence pulse and stops on a light flash. The TAC operates,
therefore, fewer times than if it started with every light flash.

When one or more emitted photons are received, a timing signal
starts the TAC which is stopped by a (delayed) signal resulting
from a lamp flash. If the gate is open, i.e., if the photocurrent
in the EF channel has the proper magnitude, the event is stored in
the proper channel of the pulse-height analyzer.

Figure 4 illustrates energy windowing. The solid curve was
taken at an efficiency of 1% with no energy windowing. After the
photon flux was raised so that the collection efficiency was 50%,
the data were biased toward shorter times and appeared as the dashed
curve. With the photon flux fixed, data were again taken, but this
time with energy windowing and an efficiency of 28%. The crosses
are representative samples of these data. Analyses of the data
yielded 8.11, 6.70, and 8.06 nsec for the three cases.

F.   Noise in a Monophoton Fluorometer

We shall define noise very broadly as anything that changes the
data from a perfect decay curve. We shall identify the following
sources of noise: photomultiplier distortion, radiofrequency pickup,
dark current, and statistical fluctuations.

1. Photomultiplier Distortion
Photomultipliers are far from ideal devices. Upon receipt of

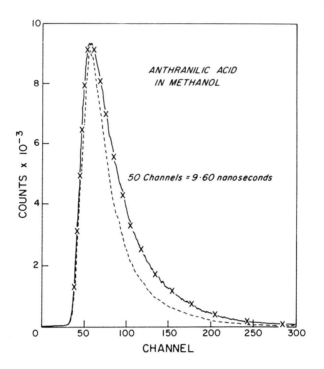

FIG. 4. Demonstration of energy windowing. The sample was an-
thranilic acid in ethanol. The solid curve (—) was taken at 1%
efficiency with no energy windowing. Again, with no energy windowing, the flux was raised so that the efficiency was 50%. The emission measured is shown as the dashed curve (---), biased toward
shorter times. The photon flux was then held constant, but energy
windowing was used. The efficiency was now 28%. The crosses show
representative data points. The low efficiency decay was 8.11 nsec,
and the measured decay using energy windowing was 8.06 nsec.

a photon, the output photocurrent may not be sharp. The single

electron response is a broad function of time. Autocorrelation

measurements even show periodicities [23].

Let $D(t)$ = output if a photoelectron is produced at $t = 0$.

$D(t)$ will be called the distortion function. We now assume that

the pulse-height analyzer, and associated electronics form a system

that is stationary in time, i.e., if a photoelectron is produced at

time u instead of at time zero, we assume that the output is $D(t - u)$.

Let F_T (= F_{True}) be the distribution in time of the photoelectrons and F_M (= $F_{Measured}$) be the measured, distorted curve.

In a monophoton experiment the data are collected by the simple addition of elementary events. We may therefore write

$$F_M = \int_{-\infty}^{\infty} D(t - u)F_T(u)du \qquad (1)$$

F_M is the convolution of the distortion function and the photoelectron distribution. We may write this as

$$F_M = D * F_T \qquad (2)$$

because we shall want to make use of the associative law of convolutions which states that

$$a * (b * c) = (a * b) * c \qquad (3)$$

The true emission F_T is a convolution of the true excitation E_T, and the response to a δ function excitation $f(t)$.

$$F_T = \int_{-\infty}^{\infty} E_T(t - u)f(u)du \qquad (4)$$

Since we are considering a lamp <u>flash</u>, we can always set our zero of time before the flash starts so that

$$E_T(t) = 0 \quad \text{if} \quad t < 0 \qquad (5)$$

or, what is equivalent,

$$E_T(t - u) = 0 \quad \text{if} \quad u > t \qquad (6)$$

Equation (4) becomes

$$F_T = \int_{-\infty}^{t} E_T(t - u)f(u)du \qquad (7)$$

Also, since $f(u)$ is the response to a δ function at $t = 0$,

$$f(u) = 0 \quad \text{if} \quad u < 0 \qquad (8)$$

because the response cannot precede the excitation.

$$\therefore \quad F_T = \int_0^t E_T(t - u)f(u)du \tag{9}$$

In the literature of fluorescence decay, Eq. (9) is the more familiar form, but (4) is equivalent to it under the conditions of (5) and (8).

Equation (4) may also be written as

$$F_T = E_T * f \tag{10}$$

and

$$D * F_T = D * (E_T * f) = (D * E_T) * f \tag{11}$$

or

$$F_M = E_M * f \tag{12}$$

$$F_M(t) = \int_{-\infty}^{\infty} E_M(t - u)f(u)du \tag{13}$$

Again, we may always choose a zero of time so that

$$E_M(t) = 0 \quad \text{for} \quad t < 0 \tag{14}$$

Consequently, we may write

$$F_M(t) = \int_0^t E_M(t - u)f(u)du \tag{15}$$

Equation (15) has exactly the form of Eq. (9).

The object of any time decay experiment is to find the significant decay parameters contained in f. Thus, it may be seen that the measured excitation and fluorescence may be used as if they were the true excitation and fluorescence. The ability to do this is one of the most important advantages of the monophoton procedure. However, two conditions are needed: The distortion function for excitation must be the same as the distortion function for emission, and the distortion function must not vary across the measuring time span of the instrument.

The first condition implies that the same photomultiplier must be used to measure E and F, and furthermore, that the conditions of measurement, for example the geometry, must be the same. It is meaningless, of course, to speak of two geometries as being exactly the same. Once F has been measured, to measure E, the sample is customarily simply replaced by a scattering sample. Whether this is good enough for a particular measurement is not, however, something that can be determined from a priori reasoning alone.

A similar statement may be made about the second requirement. The distortion function might vary across the measuring time span because, for example, the pulse-height analyzer was not linear. No instrument is perfectly linear, and, at least at the moment, there is no way yet known for calculating how much nonlinearity is tolerable for a particular type of measurement.

There are a number of implications. On the one hand, meaningful measurements are by no means ruled out by photomultiplier distortions. In fact, as was just seen, even very large distortions may be tolerated under appropriate conditions. On the other hand, for high-precision data we do not yet know how to calculate how much the distortion functions for E and F may differ, nor do we know how much nonlinearity in the pulse-height analyzer, and associated circuitry, is tolerable. It is therefore imperative to devise experimental procedures for testing an instrument's quality in any particular type of experiment. Certain tests that we have found useful will be discussed below.

2. Radiofrequency Pickup

Radiofrequency (rf) pickup is the curse of monophoton fluorometry. Pulses arising from the lamp and/or the pulsing circuit may radiate through the air to be picked up by various parts of the circuitry. Alternately, they may travel via various leads to parts of the measuring circuit. There, they may be reflected at junctures either once or a number of times.

Severe rf pickup may be seen visually as a set of bumps super-imposed on the decay curve. Figure 5 shows a particularly horrible example.

To eliminate rf pickup is often a tedious cut-and-try process. It is well to shield the lamp and pulsing circuit by a copper box. We have routinely used only double-shielded cables, and we have carefully terminated each cable. We have found that, for our in-strument at least, we cannot use lamp pulsing voltages above 4000 V. However, no definite recipe for eliminating rf troubles can be given. Such difficulties depend, in fact, on the precise geometrical ar-rangement of each assembly and on the environment of the instrument. However, it is possible to eliminate rf trouble, and Fig. 6 shows data taken in its absence.

3. Dark Current

In the monophoton procedure, dark current is a severe problem

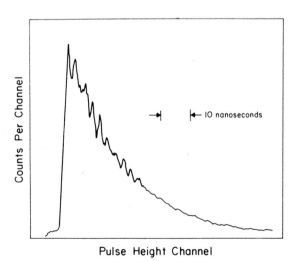

FIG. 5. Emission with a particularly horrible example of dis-tortion due to rf pickup.

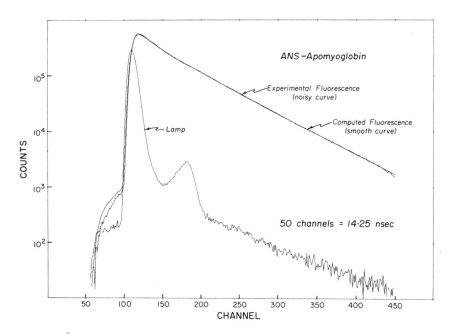

FIG. 6. Example of smooth emission data obtained after the elimination of rf pickup. The sample was apomyoglobin tagged with 1-
anilino-8-naphthalene sulfonate. The run is discussed in Schuyler
and Isenberg [15].

only if the attenuation method is used to obtain monophotons, and
then only if the precision needed for an analysis requires that the
efficiency be kept extremely low. In that case the dark current
may become competitive with the photocurrent.

With energy windowing the efficiency is usually high enough so
that dark current is not a severe problem. We find that ordinary
good photomultiplier technique will keep the background down to
about one count per second. At the present time, our counting rate
is usually in the range of 3 to 4 kHz. We generally reserve about
50-75 channels of our pulse-height analyzer for a measurement of
background. The first step in data analysis is then a computation
of the mean background, followed by a subtraction of the background
from the counts in each channel.

4. Statistical Fluctuations

The number of counts in each channel fluctuates about a mean. The true value is only estimated by the counts in a given channel. We shall call the error made owing to the finite number of counts the counting error. It is of some importance to give a theoretical estimate of the counting error, and Mullooly [24] has made such a study when the data are analyzed by the method of moments.

Mullooly's work will not be reviewed in detail here. A few points may, however, be noted. Mullooly has given implicit relationships which enable one to calculate the variances of the time constants, and amplitudes of decay, from experimental data. The basic assumption is that the probability of getting N1 counts in channel 1, N2 in channel 2, etc., is given by a multinomial probability law, i.e., that the only error is a counting error.

It is therefore of interest that we find that, generally, our analyses are good when Mullooly's equations predict a low variance and are bad otherwise. Thus, at least for such cases, the implication is that the dominant noise is counting error. It should be emphasized, however, that as we increase the number of counts in a given experiment, we should approach a point where a noncounting source of noise determines the error. Beyond this point statistical analyses will no longer yield a valid estimate of the error. It is therefore not sufficient to rely only upon a statistical estimate of the error in any given case. An experimental criterion is needed.

For low numbers of counts, the analysis of data should be bad; for very high counts it should be good, at least as far as counting error goes. Therefore, in the low, or inadequate, range the calculated parameters will vary with the number of counts; in the high, or good, range, it will not. As more and more counts are taken, however, the point will be reached where drift in the lamp or electronics will limit the accuracy. Consequently, too many or too few counts are both bad.

Experimentally, we find that the above expectations are borne out (Table 1). It is therefore necessary to determine that the

TABLE 1

Variation of Parameters with Counts.
Sample: 7 x 10^{-5} M Histone IV in 0.005 M Cacodylate Buffer;
pH 6.2 Plus 5 x 10^{-6} M N-Methyl-2-Anilino-6-Naphthalene Sulfonate

Counts in peak channel x 10^{-3}	α_1/α_2	τ_1, nsec	τ_2, nsec
2.7	0.14	19.5	6.9
5.2	0.21	16.6	5.1
7.6	0.22	16.2	4.9
11.4	0.24	15.7	4.7
20.9	0.24	15.6	4.8
35.9	0.24	15.8	5.0

counting range is a good one for any particular type of experiment. This may be done in a direct way. The requirement is that the data not vary as the precision is changed. Two or more runs may be taken at appreciably different counts, say a factor of 2 or 3. To be acceptable experiments, the computed parameters of these experiments must agree with one another.

Statistical expressions are important aids, but they cannot be substitutes for experimental tests of the validity of derived parameters.

III. ANALYSIS OF DATA

The object of any data analysis is to obtain the physical parameters in f, the response to a δ function flash. The object is not how to fit a curve to a given set of data. Figure 1 shows that visual observation is inadequate. Since, however, visual observation is qualitative, one could hope to find some statistical measure,

for example, least mean square, that might be able to specify when
a fit is good. Grinvald and Steiner [24a] have shown that a simple
mean-square measure is not adequate to fit time-decay data. However, they have given examples that illustrate that a weighted mean-
square measure may serve as a method of fitting, provided the weighting is taken to be proportional to the noise-to-signal ratio.

It is sometimes thought that the entire difficulty of analyzing
decay data stems from the necessity of deconvoluting Eq. (15). This
is not true. Even if E_M were a δ function so that $F_M = f$, difficulties remain. For example, f is often a sum of exponentials. To determine the parameters in a sum of exponentials from experimental
data is a difficult, ill-conditioned problem [25,26]. When, in
addition, the finite breadth of the exciting lamp cannot be neglected,
we are also faced with deconvoluting Eq. (15).

In principle, formalisms exist for a direct deconvolution. The
Fourier transform, or the Laplace transform, of a convolution is the
product of the transforms of the individual functions. Thus, if we
let $\mathcal{L}(F)$ be the Laplace transform of F and $\mathcal{L}^{-1}$ be the inverse transform, we may formally write

$$f(u) = \mathcal{L}^{-1} \frac{\mathcal{L}(F_M)}{\mathcal{L}(E_M)} \tag{16}$$

Since F_M and E_M are data, Eq. (16) describes a formal solution.
However, deconvolution by transforms appears to be extremely sensitive to noise [27], and a method of circumventing the noise problem
is still missing.

At least at the present time, it appears to be easier to assume
a functional form for f as a first step in the analysis. This, of
course, limits the method to problems having that functional form,
or problems that can be approximated by that functional form. There
are some problems for which a theoretically significant functional
form is not yet known. An important example is the study of time-
dependent emission spectra [4,28,29]. It is unfortunate, therefore,
that there is as yet no practical, general deconvolution scheme.

In this chapter, f will be restricted to a sum of exponentials. This limits the procedure, but nevertheless a sum of exponentials covers a large number of important cases. The decay of the anisotropy of emission from a rigid body is a sum of exponentials [30], as is, of course, the decay of the total emission from a heterogeneous population. Furthermore, at the present time, much more is known about such an analysis than about any other form.

Our laboratory uses the method of moments [19,31] routinely to analyze decay data under the assumption that f is a sum of exponentials. We have published examples of such analyses [15,32,32a] and others are given here or are in press.

We shall therefore write

$$f = \sum_{i=1}^{N} \alpha_i e^{-t/\tau_i} \tag{17}$$

N will be called the number of components, α_i the amplitude of the ith component, and τ_i the decay time, or time constant of the ith component.

We now restate the problem. From the data we must determine N, α_i, and τ_i.

The method of moments for time decay studies originated with Bay [31]. Brody [33] used it for a one-component analysis, and Wahl and Lami [34] used a combination method of moments and subtraction procedure.

Basically, the method of moments substitutes moments of the data for the data themselves.

Let

$$\mu_k = \int_0^\infty t^k F(t)\ dt \tag{18}$$

$$m_k = \int_0^\infty t^k E(t)\ dt \tag{19}$$

μ_k and m_k are the kth moments of the fluorescence and excitation, respectively. We shall, for simplicity, drop the subscripts in Eq. (15) and write it as

$$F(t) = \int_0^t E(t - u)f(u) \, du \qquad (20)$$

From (18), (19), and (20) one may show [19] that

$$\frac{\mu_k}{k!} = \sum_{s=1}^{k+1} G_s \frac{m_{k+1-s}}{(k + 1 - s)!} \qquad (21)$$

where

$$G_s = \sum_{i=1}^{N} \alpha_i \tau_i^s \qquad (22)$$

For successive values of k, Eq. (21) provides a set of equations:

$$\mu_0 = G_1 m_0 \qquad (23a)$$

$$\mu_1 = G_1 m_1 + G_2 m_0 \qquad (23b)$$

$$\frac{\mu_2}{2!} = \frac{G_1 m_2}{2!} + G_2 m_1 + G_3 m_0 \qquad (23c)$$

$$\cdot \qquad \cdot$$
$$\cdot \qquad \cdot$$
$$\cdot \qquad \cdot$$

These equations are linear in G_1, G_2, ... If the moments μ_k and m_k are known, G_s may be determined for successively higher values of s.

If N components exist, then G_1, G_2, ..., G_N, G_{N+1}, ..., G_{2N} are completely equivalent in information content to α_1, α_2, ..., α_N, τ_1, τ_2, ..., τ_N. Given either set, one may determine the other.

To calculate the decay parameters from G_1, ..., G_{2N}, one proceeds as follows. We form a polynomial in x of the Nth degree

$$P_N(x) = \begin{vmatrix} 1 & x & x^2 & \cdots & x^N \\ G_1 & G_2 & G_3 & \cdots & G_{N+1} \\ G_2 & G_3 & G_4 & \cdots & G_{N+2} \\ \vdots & & & & \\ G_N & G_{N+1} & G_{N+2} & & G_{2N} \end{vmatrix}$$

The N zeros of the polynomial are $\tau_1, \tau_2, \ldots, \tau_N$.

Properties of polynomials of the type of P_N are well-known [35]. An elementary proof of the assertion that the zeros of P_N are the time constants is as follows.

It may be seen that

$$P_N(x) = \begin{vmatrix} 1 & 0 & 0 & \cdots & 0 \\ 0 & \alpha_1\tau_1 & \alpha_2\tau_2 & \cdots & \alpha_N\tau_N \\ 0 & \alpha_1\tau_1^2 & \alpha_2\tau_2^2 & \cdots & \alpha_N\tau_N^2 \\ \vdots & & & & \\ 0 & \alpha_1\tau_1^N & \alpha_2\tau_2^N & \cdots & \alpha_N\tau_N^N \end{vmatrix} \begin{vmatrix} 1 & x & x^2 & \cdots & x^N \\ 1 & \tau_1 & \tau_1^2 & \cdots & \tau_1^N \\ 1 & \tau_2 & \tau_1^2 & \cdots & \tau_2^N \\ \vdots & & & & \\ 1 & \tau_N & \tau_N^2 & \cdots & \tau_N^N \end{vmatrix}$$

The second determinant vanishes whenever x equals a decay constant, for then two rows are equal. Therefore, $x = \tau_1, \tau_2, \ldots, \tau_N$ are solutions of

$$P_N(x) = 0 \tag{25}$$

Since P_N is a polynomial of the Nth degree, these are the only solutions.

Once the time constants are known, the amplitudes are easily obtained:

$$G_1 = \sum_{i=1}^{N} \alpha_i \tau_i \tag{26a}$$

$$G_2 = \sum_{i=1}^{N} \alpha_i \tau_i^2$$

.
.
.

(26b)

$$G_{2N} = \sum_{i=1}^{N} \alpha_i \tau_i^{2N}$$

Equations (26) provide a linear set which can be solved for $\alpha_1, \alpha_2, \ldots, \alpha_N$.

In principle, then, the problem is solved. In practice, however, one is faced with three problems, one that is peculiar to the method of moments, and two that are completely general and arise in any method.

We begin by a discussion of the problem that is specific to the method of moments. Equation (21) is valid only if the moments defined by Eqs. (18) and (19) go to infinity. We may ask if the moments may be approximated by

$$\mu_k \cong \int_0^T t^k F(t) \, dt \tag{27}$$

$$m_k \cong \int_0^T \iota^k E(t) \, dt \tag{28}$$

for experimentally reasonable values of T.

$E(t)$ has a short enough breadth so that Eq. (28) can usually be used in place of (19). On the other hand, for reasonable values of T, Eq. (27) is generally not a good approximation [19]. Furthermore, the higher the moment, the worse the approximation. Consequently the problem, which may be called the cut-off problem, becomes more serious as N becomes larger.

The first approach to a solution of the cut-off problem was to use an iterative procedure. With such a method, one first analyzes the data, using Eq. (27) in place of (19). One then makes an approximate correction to the moments by adding

$$\Delta\mu_k = \int_T^\infty t^k \sum_{i=1}^N \alpha_i e^{-t/\tau_i} \, dt \qquad (29)$$

where the α_i and τ_i are the (incorrect) parameters obtained in this first trial.

The analysis is repeated, and a new set of parameters is obtained to make a new correction. This iterative procedure is continued until self-consistency is attained.

For one or two components, iterative looping works well. We give an example in Table 2. 2.7×10^{-4} carbazole in 95% ethanol has a decay time of 9.6 nsec. 4.7×10^{-5} M anthracene in 95% ethanol has a decay time of 4.6 nsec. Together, they yield a decay curve which, upon a two-component analysis, converged in thirteen loops to $\tau_1 = 9.9$ nsec and $\tau_2 = 4.5$ nsec. Thirteen loops is not excessive and, in fact, is the order of magnitude found for most two-component analyses we have made.

If, however, we now attempt a three-component analysis, we meet excessive looping. Table 2 illustrates that even after 500 loops, there was still no convergence. We postpone for the moment a discussion of why one might want to analyze for three components in such a case. We point out, however, that aside from general considerations of testing the system, one might ask here, for example, if the solutes interacted and, if so, if they thereby showed one or more additional components.

Convergence is aided enormously by a procedure that we have called exponential depression [32a]. Exponential depression involves a simple transformation.

If we multiply Eq. (20) by $e^{-\lambda t}$, where λ is some fixed positive number, we obtain

$$e^{-\lambda t} F(t) = \int_0^t e^{-\lambda(t-u)} E(t-u) e^{-\lambda u} f(u) \, du$$

TABLE 2

Use of Exponential Depression in Aiding the
Convergence of Iterative Looping, and Determination of
the Number of Components by Incrementation of N. Sample:
2.7 x 10^{-4} M Carbazole and 4.7 x 10^{-5} M Anthracene in 95% Ethanol

Analysis	No. of Loops	α_1	τ_1	α_2	τ_2	α_3	τ_3
Two-component	13	0.030	9.87	0.071	4.47		
Three-component	100	0.037	9.03	0.064	4.16	0.0009	16.3
No depression	500	0.037	9.24	0.065	4.19	0.0002	23.6
			No convergence				
Three-component exp. dep. $F_\lambda(T)/F(T) = 0.01$	6	0.032	9.72	0.070	4.32	3.4 x 10^{-6}	-5.77

Now define

$$F_\lambda(t) = e^{-\lambda t} F(t) \tag{30}$$

$$E_\lambda(t) = e^{-\lambda t} E(t) \tag{31}$$

$$f_\lambda(t) = e^{-\lambda t} f(t) \tag{32}$$

Then,

$$F_\lambda(t) = \int_0^t E_\lambda(t - u) f_\lambda(u)\, du \tag{33}$$

In other words, the convolution is invariant to the transformation (30)-(32).

These equations depress the data exponentially. For long times $F_\lambda(t)$ becomes much smaller than $F(t)$, and, as a consequence, the cut-off correction is sharply reduced.

Table 2 illustrates that, for a depression for which $F_\lambda(T)/F(T)$ = 0.01, a three-component analysis yielded satisfactory convergence in only six loops, even though there was no convergence even at 500. Furthermore, it should be stated that not only was there no convergence at 500 loops, but the parameters were changing slowly, and it was clear that the analysis was far from the convergence point.

We now turn to the second of our problems. We wish to determine the number of components. This problem is, of course, by no means limited to the method of moments. It will arise no matter what type of analysis is used.

The procedure that we have used to find N is also independent of the use of the method of moments. It can be applied to any method. We simply analyze for successively increasing numbers of components. We stop when we see that an analysis for an additional component does not change the results significantly, but only adds a component with a relatively small amplitude.

Table 3 gives an example. It may be seen that a one-component analysis yielded

$$f = 8.32 \times 10^{-3} e^{-t/27.45}$$

while a two-component analysis yielded

$$f = 8.32 \times 10^{-3} e^{-t/27.45} + 5.7 \times 10^{-10} e^{+t/13.7}$$

Note that the amplitude of the second component is negligibly small compared to the first, while the first component is the same in the two-component analysis as it was in the one-component analysis. We therefore have reason for accepting a one-component analysis of the data. (The negative time constant for the second component is not significant, since the component is really nonexistent. Its appearance is due only to noise.)

TABLE 3

Analysis of Decay of Emission from
1.8×10^{-5} M E-ATP in 0.005 M Cacodylate Buffer pH 7.0

Analysis	α_1	τ_1	α_2	τ_2
One-component	8.32×10^{-3}	27.45		
Two-component	8.32×10^{-3}	27.45	5.7×10^{-10}	-13.7

Another example is shown in Table 2. It may be seen that a
three-component analysis yielded parameters that were the same as a
two-component analysis, plus the addition of a third having a negligible amplitude.

The last problem facing an investigator is to determine if the
data are good enough to be analyzed. As with any instrument, a new
assembly requires testing and proof of performances. Such tests
will vary from one laboratory to another, and we have presented
examples run on our instrument [15]. However, even after such tests
are given, every measurement will require a demonstration that at
least the counting error is not too large. A useful criterion is
to make independent runs at different numbers of counts and test to
see if closely similar decay parameters are obtained [15,19,32,36].

ADDENDUM

A technique named moment index displacement (MD) has been introduced into the method of moments [36]. An MD of order n is the
process of calculating ν values of τ_i and ν values of a_i from the
set $(G_{n+1}, G_{n+2}, \ldots, G_{n+2\nu})$ instead of $(G_1, G_2, \ldots, G_{2\nu})$. MD will
correct for several instrumental errors. An MD of any order will

completely correct for any size of light-scattering error. MD will also correct for small discrepancies between the measured zero point of E(t) and F(t), and it will correct for slow drifts in E(t) that occur during a measurement of F(t).

Closed form expressions have been given [37] for the statistical errors in measurement of decay times when the method of moments is used to analyze the data.

ACKNOWLEDGMENTS

The work reported in this chapter was supported by Public Health Service Grant CA 10872, National Science Foundation Grant GB-33498, and a grant from the Oregon Branch of the American Cancer Society. The author thanks Dr. Robert Schuyler, Dr. Robert Dyson, Dr. Enoch Small, and Mr. Richard Hanson for their contributions to various phases of the work described here.

REFERENCES

1.   W. R. Bennett, in Advanced Quantum Electronics (J. Singer, ed.), Columbia University Press, New York, 1961.

2.   L. M. Bollinger and G. E. Thomas, Rev. Sci. Instr., 32, 1044 (1961).

3.   Y. Koechlin, Acad. Sci. Paris Compt. Rend., 252, 391 (1961).

4.   W. R. Ware, in Creation and Detection of the Excited State, Vol. 1, Part A, Marcel Dekker, Inc., New York, 1971.

5.   R. D. Spencer and G. Weber, Ann. New York Acad. Sci., 158, 361 (1969).

6.   J. H. Malmberg, Rev. Sci. Instr., 28, 1027 (1957).

7.   L. Hundley, T. Coburn, E. Garwin, and L. Stryer, Rev. Sci. Instr., 38, 488 (1967).

8.   I. B. Berlman, O. J. Sterngraber, and M. J. Benson, Rev. Sci. Instr., 39, 54 (1968).

9.   W. Ware, University of Minnesota, Tech. Rept. No. 3 on U.S. Office of Naval Res. Contract N0014-67-A-0113-0006 (1969).

10. J. T. D'Alessio, P. K. Ludwig, and M. Burton, Rev. Sci. Instr., 35, 1015 (1964).

11. J. Yguerabide, Rev. Sci. Instr., 36, 1734 (1965).

12. J. T. D'Alessio and H. Lanza, Rev. Sci. Instr., 39, 1029 (1968).

13. J. G. Innes and O. A. Kerns, U.S. Atomic Energy Commission (UCRL-9726) (1961).

14. T. Tao, Biopolymers, 8, 609 (1969).

15. R. Schuyler and I. Isenberg, Rev. Sci. Instr., 42, 813 (1971).

16. P. B. Coates, J. Physics D. Applied Phys., 3, 1290 (1970).

17. D. A. Gedcke and W. J. McDonald, Nucl. Instr. Methods, 55, 337 (1967).

18. D. A. Gedcke and W. J. McDonald, Nucl. Instr. Methods, 58, 253 (1968).

19. I. Isenberg and R. D. Dyson, Biophys. J., 9, 1337 (1969).

20. P. B. Coates, J. Physics E., 1, 878 (1968).

21. C. C. Davis and T. A. King, J. Phys. A., 3, 101 (1970).

22. R. E. Simon, A. H. Sommer, J. J. Tietjen, and B. F. Williams, Appl. Phys. Lett., 13, 355 (1968).

23. M. Corti and A. Vendramini, Rev. Sci. Instr., 42, 1300 (1971).

24. J. Mullooly, Biophys. J., 13, 1109 (1973).

24a. A. Grinwald and I. Z. Steinberg, Anal. Biochem., 59, 583 (1974).

25. C. Lanczos, Applied Analysis, Prentice Hall, New Jersey, 1956.

26. R. W. Hamming, Numerical Methods for Scientists and Engineers, McGraw-Hill, New York, 1962.

27.   I. H. Munro and I. A. Ramsey, J. Sci. Instr., 1, 147 (1968).

28.   W. R. Ware, S. K. Lee, and P. Chow, Chem. Phys. Lett., 2, 356 (1968).

28a. R. Schuyler, R. D. Dyson, and I. Isenberg, Photochem. Photobiol., 15, 395 (1972).

29.   L. Brand and J. R. Gohlke, J. Biol. Chem., 246, 2317 (1971).

30.   G. G. Belford, R. L. Belford, and G. Weber, Proc. Natl. Acad. Sci., 69, 1392 (1972).

31.   Z. Bay, Phys. Rev., 77, 419 (1950).

32.   I. Isenberg, R. D. Dyson, and R. Hanson, Biophys. J., 13, 1090 (1973).

33.   S. S. Brody, Rev. Sci. Instr., 28, 1021 (1957).

34.   P. Wahl and H. Lami, Biochem. Biophys. Acta, 133, 233 (1967).

35.   N. I. Ahiezer and M. Krein, Some Questions in the Theory of Moments, Amer. Math. Soc., Providence, R. I., 1962.

36.   I. Isenberg, J. Chem. Phys., 59, 5696 (1973).

37.   I. Isenberg, J. Chem. Phys., 59, 5708 (1973).

Chapter 3

FLUORESCENCE POLARIZATION: SOME TRENDS AND PROBLEMS

Izchak Z. Steinberg
Department of Chemical Physics
The Weizmann Institute of Science
Rehovot, Israel

I. INTRODUCTION

In the interaction of light and matter, the relation of the state of polarization of the light and the geometrical arrangement in space of the absorbing or emitting molecule plays a major role. The absorption of linearly polarized light by oriented arrays of molecules in crystals or in stretched polymer films depends on the direction of polarization of the beam. Light which is partly linearly polarized is emitted under proper conditions by molecules embedded in rigid media. Right- and left-handed circularly polarized light is absorbed or emitted with different intensity by chiral molecules in solution, or by nonchiral molecules placed in a magnetic field. As early as 1911 Selényi [1] illustrated experimentally that the distribution in space of the emission from fluorescein was characteristic of an electric dipole oscillator. By similar procedures, the radiation field due to one of the emission bands of europium ion was shown to be characteristic of a magnetic dipole oscillator [2]. By precise positioning of chromophores relative to other chromophores or relative to reflecting surfaces, by use of monolayer assemblies, Kuhn and co-workers were able to study the multipolar nature of absorbing and emitting molecules in a number of elegant ways [3].

Theory did not lag behind. The general principles of light absorption and emission processes have become textbook knowledge, and the polarization characteristics of transitions between quantum states has been extensively discussed. The theoretical developments led the way to numerous applications. Thus, the study of the absorption or emission of linearly polarized light by anisotropic systems has become an important tool in the assignment of electronic transitions and in the verification of quantum mechanical calculations concerning the molecules involved. Similarly, the interaction of circularly polarized light with chiral systems has established itself as a major and an indispensable instrument for the understanding of such systems in chemistry and biology.

Probing molecules with linearly polarized light or with circularly polarized light yields information which is complementary
rather than redundant. Absorption or emission of linearly polarized
light is usually applied to the study of the vectorial behavior of
electric dipole transition moments, while the difference in interaction of isotropic media with circularly polarized light of opposite
sense is due to a combination of the relevant electric and magnetic
transition dipole moments of the absorbing or luminescent molecules,
and is related to the molecular conformation. Until recently very
little activity has been going on in the study of the circular polarization of emitted light in luminescence processes. This is particularly striking in view of the fact that circular dichroism, i.e.,
the difference in absorption of right- and left-handed circularly
polarized light, is an active and busy field, especially in organic
chemistry and in biochemistry. The technical difficulties involved
in the measurement of the circular polarization of luminescence may
be responsible, to a large extent, for the neglect of this field.
With recent advances in instrumentation, the possibilities offered
by the study of the circular polarization of fluorescence and phosphorescence will hopefully be explored and exploited.

In contradistinction to the study of the circular polarization
of luminescence, the investigation and application of the linear
polarization of emitted light has flourished in biochemistry. It
has been used extensively to measure rotational diffusion coefficients, which have important implications regarding the size and shape
of macromolecules [4,5]. Furthermore, it has yielded invaluable
information regarding the flexibility of segments of macromolecules
and the rigidity of the binding of ligands to them [6]. Linear polarization of fluorescence has also served as the principal tool for
the study of transfer of electronic excitation energy between chromophores of the same kind [7].

The subject of the polarization of fluorescence is very wide
indeed, and a review of the field is not attempted here. The scope
of the present article is necessarily rather limited. Some basic

concepts concerned with the absorption and emission of polarized
light will be examined and a few implications discussed. An attempt will be made to describe a few of the new trends in the field.

II. BASIC CONSIDERATIONS

A. Mixed Polarizations in Electronic Bands

When dealing with fluorescence phenomena, it is common practice
for chemists and biochemists to switch back and forth between classical and quantum mechanical considerations according to convenience.
The quantum mechanical picture is adopted when energy, frequency,
Stokes' shifts, relaxation processes in the excited state, etc., are
considered. When the absorption or emission of linearly polarized
light is discussed, the visualization of the absorbing or emitting
molecules as classical oscillators is very common. Furthermore, a
single oscillator is often attributed to an emission band or an absorption band involving one electric transition. The fact that this
approach is quite satisfactory in many cases indicates that it is a
good approximation. One should be prepared for surprises, however,
if the classical analogy is carried out indiscriminately. The problem of the absorption or emission of polarized light can, and should,
be treated rigorously by quantum mechanical theory. Fundamental
considerations of the interaction of light with complex molecules
were presented by Herzberg and Teller [8,9]. Albrecht has extended
this treatment with special reference to the linear polarization of
allowed transitions and to the method of photoselection, and has
illustrated the theoretical study with some interesting experimental
results [10-12]. The vibronic details of circular dichroism spectra
were studied theoretically by Moffitt and Moscowitz along similar
lines [13].

It may be of some value to list some of the key points concerning the character of the polarization of absorption and emission processes. (1) For a transition between well-defined quantum states, the molecule behaves as a multipolar oscillator. (2) The electric dipole term predominates for allowed transitions. The magnetic dipole transition moment assumes, however, importance in the study of the interaction of chiral molecules with circularly polarized light. (3) The wave function of a quantum state of a molecule may be approximated as a product of an electronic wave function and a wave function of the nuclei (Born-Oppenheimer approximation). These states are commonly called <u>vibronic</u> states. It should be noted that the electronic wave function depends on the position of the nuclei. (4) Transitions accompanied by absorption or emission of light can take place between vibronic levels involving different states, subject to certain selection rules. For transitions between a pair of electronic states involving different vibrational levels, the intensities will be different as a rule (Franck-Condon principle). (5) The common notion that transitions between a pair of electronic states involving different vibrational levels have the same polarization properties holds to a first approximation. This approximation may, however, be seriously inadequate. (6) To a similar level of approximation, the circular dichroism spectrum of a compound should have the same shape as its absorption spectrum for each electronic band. Again, this approximation may be inadequate.

The statement that the transition dipole moment may not have a unique direction for transitions involving a single pair of electronic levels may come as a surprise to some. As a matter of fact, changes in polarization across an absorption band have been commonly interpreted as conclusive evidence for overlap of different electronic transitions in the band. A brief review of the formal background of this important point may therefore be of interest.

Within the limit of the Born-Oppenheimer approximation, the wave function Φ_{kj} of a molecule in the jth vibrational level of the kth electronic state may be written as a product of the corresponding electronic and nuclear wave functions, θ_k and φ_j^k, respectively.

$$\Phi_{kj} = \theta_k(x,Q)\varphi_j^k(Q) \tag{1}$$

In Eq. (1), x and Q refer to the complete set of internal co-ordinates required to locate all the electrons and all the nuclei, respectively. The electric dipole transition moment, $\underline{M}_{gi,kj}$, for the transition from the vibronic state g-i to the vibronic state k-j is given by Eq. (2) (for simplicity, real wave functions are assumed):

$$\underline{M}_{gi,kj} = \int \Phi_{gi}[\underline{m}_e(x) + \underline{m}_e(Q)]\Phi_{kj} \, dx \, dQ$$

$$= \int \theta_g(x,Q)\underline{m}_e(x)\theta_k(x,Q)\varphi_i^g(Q)\varphi_j^k(Q) \, dx \, dQ$$

$$+ \int \theta_g(x,Q)\theta_k(x,Q)\varphi_i^g(Q)\underline{m}_e(Q)\varphi_j^k \, dx \, dQ \tag{2}$$

In Eq. (2), $\underline{m}_e(x)$ and $\underline{m}_e(Q)$ represent the electronic and nuclear terms, respectively, of the electric dipole moment operator. The second integral on the right-hand side of Eq. (2) vanishes, since integration with respect to x, for any value of Q, is identically zero owing to the orthogonality of the wave functions of the electronic states g and k. As for the first integral in Eq. (2), if θ_g and θ_k did not vary with Q in the vicinity of the equilibrium position Q_o of the nuclei, it would have assumed a relatively simple form:

$$\underline{M}_{gi,kj} = \int \theta_g(x,Q_o)\underline{m}_e(x)\theta_k(x,Q_o) \, dx \int \varphi_i^g \varphi_j^k \, dQ$$

$$= \underline{M}_{g,k}(Q_o) \cdot FCI \tag{3}$$

where

$$\underline{M}_{g,k}(Q_o) = \int \theta_g(x,Q_o)\underline{m}_e(x)\theta_k(x,Q_o) \, dx \tag{4}$$

and

$$FCI \text{ (the Franck-Condon integral)} = \int \varphi_i^g(Q)\varphi_j^k(Q) \, dQ \tag{5}$$

To this degree of approximation, the vectorial properties of the electric dipole transition moment $\underline{M}_{gi,kj}$ between any two vibronic levels belonging to one pair of electronic states g and k are determined by a single vector $\underline{M}_{g,k}(Q_o)$ which is common to all of these vibronic transitions.

This approximation, which neglects the dependence of $\theta_g(x,Q)$ and $\theta_k(x,Q)$ on the nuclear coordinates Q, fails completely for forbidden transitions for which $\underline{M}_{g,k}(x,Q_o)$ is zero (e.g., because of symmetry properties of the relevant wave functions). Such transitions may still be observed for appropriate nuclear vibrational levels due to the dependence of the electronic wave functions on the coordinates of the nuclei. The treatment first proposed by Herzberg and Teller for forbidden transitions was shown by Albrecht to be pertinent to, and sometimes quite important for, allowed transitions as well. Much insight into the effect of vibrations on the characteristics of electronic absorption bands (and for that matter also of emission bands) was gained by using the wave functions $\theta_s(x,Q_o)$, which are valid for the equilibrium positions of the nuclei in the ground state, as a complete set of zero-order wave functions and considering the nuclear motion as a perturbation in the electronic Schroedinger equation. Thus, for the first-order perturbation treatment,

$$\theta_k(x,Q) = \theta_k(x,Q_o) + \sum_s \lambda_{ks}(Q)\theta_s(x,Q_o) \tag{6}$$

where the λ_{ks}'s are functions of the nuclear coordinates and can be evaluated by standard expressions [10]. It need only be mentioned that among other things they are inversely proportional to $\Delta E^o_{s,k}$, the difference in energies of the s and k states at the nuclear positions Q_o. An expression similar to that of Eq. (6) applies also to the ground state electronic wave function $\theta_g(x,Q)$. However, the corresponding λ_{gs} values for the ground state are relatively small because of the relatively large corresponding ΔE^o_{gs} values and are usually neglected.

Insertion of Eq. (6) in Eq. (2) yields

$$M_{gi,kj} = M_{g,k}(Q_o) \int \varphi_j^k(Q) \varphi_i^g(Q) \, dQ$$

$$+ \sum_s M_{g,s}(Q_o) \int \varphi_j^k(Q) \lambda_{ks}(Q) \varphi_i^g(Q) \, dQ \qquad (7)$$

Hence the important conclusion that <u>different vibrational transitions within a certain electronic band need not be polarized in the same direction</u>. Symmetry considerations are helpful in determining to which vibrations the first or second terms in Eq. (7) apply [9,11]. However, for electronic bands which show poor vibrational resolution (which is often the case in biochemistry), these quantitative considerations are not of much practical application. When the vibrational transitions strongly overlap, one would observe experimentally that the polarization across a single electronic band is not constant, and that at any one wavelength the transition is not necessarily characterized by a single transition dipole moment but may be characterized by two or three nonparallel dipoles. This may occur in particular if a level θ_s exists which is close to θ_k (i.e., ΔE_{ks}^o is small) and for which M_{gs}^o is large and polarized differently than M_{gk}^o. (Refinement of the Born-Oppenheimer approximation introduces additional transitions in the spectrum. This effect has, however, been estimated to be relatively small as regards radiative transitions [9].)

    While the number of careful studies related to the polarization of different vibronic transitions within a single electronic band seems to be rather limited, the experimental studies of p-dimethoxybenzene and of N,N,N′,N′-tetramethyl-paraphenylenediamine by Albrecht show indeed that the last absorption bands of these compounds have mixed polarizations to a very marked extent (see Table III of Ref. [10]). Liptay has listed a few naphthalene derivatives which behave similarly [14]. Rhodopsin shows similar behavior [15]. The emission band of some porphyrins was found to vary markedly in polarization with wavelength [16], and the polarization spectrum is not constant for vibronic transitions involving the same electronic levels [17]. Recently it was found that the linear polarization of the fluores-

cence of dansyl [17a] and $1,N^6$-ethenoadenine-nicotinamide-dinucleo-
tide[17b] chromophores varied significantly across their emission
spectra. Moreover, the spectral behavior of the polarization
changed markedly upon binding of the chromophores to specific sites
of proteins, which is not surprising in view of the expected sensitivity of weak transitions to vibrational perturbations by the environment. The polarization spectrum across the emission band of
chromophores with a weak emission transition may thus serve as a
new parameter for probing binding sites of biopolymers and other
biological structures. In view of the above, it seems rather unfortunate that not enough vigilance has been shown to the prospects,
problems, and pitfalls involved in the possibility of occurrence of
mixed polarizations in electronic bands.

B. Absorption and Emission
of Circularly Polarized Light

As is well-known, circular dichroism (CD) carries essentially
the same information about the system studied as does optical rotatory dispersion (ORD), except that it is simpler to relate experimental data on CD than those on ORD to specific electronic transitions [18]. In the early formulations of the theory of CD, it must
have been recognized that unequal capability of absorption of light
circularly polarized in opposite senses is a closely related phenomenon to preferential emission of circularly polarized light [19].
However, until recently the study of optical rotatory power relied
exclusively on ORD and CD, and the phenomenon of circular polarization of luminescence, CPL, of chiral compounds was almost completely
neglected. It is important to realize that while the CD and CPL of
a molecule are related to each other for a transition between two
defined quantum states, the information obtained by CD and that obtained by CPL are by no means redundant. The vibronic quantum states
involved in the absorption and emission processes of complex molecules are, of course, not the same ones on the whole. Roughly

speaking, the CD of a compound is related to its chirality (and
hence conformation) in the electronic ground state, while the CPL of
the compound is related to its chirality and conformation in the
electronically excited state from which emission occurs.

In an isotropic solution of chiral molecules, the transition
probability P for excitation by circularly polarized light, per unit
light intensity, is given by [19,20]

$$P_\ell = K\{ [\int \Phi_{gi} \underline{m}_e \Phi_{kj} \, dx \, dQ]^2$$

$$+ 2 \, \mathrm{Im}[(\int \Phi_{gi} \underline{m}_e \Phi_{kj} \, dx \, dQ) \cdot (\int \Phi_{kj} \underline{m}_m \Phi_{gi} \, dx \, dQ)]\} \qquad (8)$$

$$P_r = K\{ [\int \Phi_{gi} \underline{m}_e \Phi_{kj} \, dx \, dQ]^2$$

$$- 2 \, \mathrm{Im}[(\int \Phi_{gi} \underline{m}_e \Phi_{kj} \, dx \, dQ) \cdot (\int \Phi_{kj} \underline{m}_m \Phi_{gi} \, dx \, dQ)]\} \qquad (9)$$

where the subscripts ℓ and r refer to absorption of left- and right-
handed circularly polarized light, respectively, and K is a proportionality factor. The symbol Im denotes taking the imaginary part
of the expression following it. The symbols $\underline{m}_e$ and $\underline{m}_m$ refer to the
electronic terms of the electric dipole moment operator and magnetic
dipole moment operator, respectively [19]. The nuclear terms of the
dipole moment operators have been dropped, as explained above. It
is convenient to define an anisotropy factor g_{abs} for the absorption
of light by chiral systems,

$$g_{abs} = \frac{\epsilon_\ell - \epsilon_r}{\epsilon} = \frac{\epsilon_\ell - \epsilon_r}{(\epsilon_\ell + \epsilon_r)/2} = \frac{P_\ell - P_r}{(P_\ell + P_r)/2}$$

$$= 4 \, \frac{\mathrm{Im}[(\int \Phi_{gi} \underline{m}_e \Phi_{kj} \, dx \, dQ) \cdot (\int \Phi_{kj} \underline{m}_m \Phi_{gi} \, dx \, dQ)]}{[\int \Phi_{gi} \underline{m}_e \Phi_{kj} \, dx \, dQ]^2} \qquad (10)$$

where ϵ_ℓ, ϵ_r, and ϵ are the molar extinction coefficients for left-
handed circularly polarized light, right-handed circularly polarized
light, and unpolarized light, respectively.

The probability of emission of left-handed and right-handed
circularly polarized light from a chiral molecule is proportional to

P_ℓ and P_r, respectively [Eqs. (8) and (9)]. (This statement involves the subtle assumption, which can be justified, that the arguments of Einstein for the relation between the transition probabilities of absorption and of spontaneous emission hold separately for left- and right-handed circularly polarized light.) Let us define an anisotropy factor g_{em} for the emission of light by chiral systems,

$$g_{em} = \frac{\Delta f}{f/2} = \frac{P_\ell - P_r}{(P_\ell + P_r)/2}$$

$$= 4 \, \frac{\text{Im}[\, (\int \Phi_{gi} \, \underline{m}_e \Phi_{kj} \, dx \, dQ) \cdot (\int \Phi_{kj} \, \underline{m}_m \Phi_{gi} \, dx \, dQ)\,]}{[\int \Phi_{gi} \, \underline{m}_e \Phi_{kj} \, dx \, dQ]^2} \qquad (11)$$

where Δf represents the intensity of the circularly polarized component of the light emitted from an isotropic collection of luminescent molecules (when this component is left-handed, Δf is defined as positive), and f is the total intensity of the emitted light. At first sight the expressions for g_{abs} and g_{em} seem equivalent for spectral bands involving the electronic states θ_g and θ_k, but they are not! g_{abs} and g_{em} do not, on the whole, involve the same vibrational transitions, and the wave functions which appear in these expressions depend explicitly on Q, the coordinates of the positions of the nuclei of the molecule (and possibly of surrounding solvent molecules as well). If the molecule has different conformations, or if its interactions with the environment are different, in the ground and the excited states, g_{abs} and g_{em} may assume completely different values.

The dependence of the electronic wave functions on Q can, of course, be treated by the Herzberg-Teller approach. However, to illustrate the significance of the difference between g_{abs} and g_{em}, let us assume a first approximation for $\theta_g(x,Q)$ and $\theta_k(x,Q)$. Analogously to obtaining Eq. (3), one thus obtains

$$g_{abs} = 4 \, \frac{\text{Im}\{\int \theta_g(x,Q_o{}^g) \underline{m}_e \theta_k(x,Q_o{}^g) \, dx \cdot \int \theta_k(x,Q_o{}^g) \underline{m}_m \theta_g(x,Q_o{}^g) \, dx\}}{\{\int \theta_g(x,Q_o{}^g) \underline{m}_e \theta_k(x,Q_o{}^g) \, dx\}^2} \qquad (12)$$

$$g_{em} = 4 \frac{\text{Im}\{\int \theta_g(x,Q_o^k)\underline{m}_e\theta_k(x,Q_o^k) \, dx \cdot \int \theta_k(x,Q_o^k)\underline{m}_m\theta_g(x,Q_o^k) \, dx\}}{\{\int \theta_g(x,Q_o^k)\underline{m}_e\theta_k(x,Q_o^k) \, dx\}^2} \tag{13}$$

where $Q_o{}^g$ and $Q_o{}^k$ are the equilibrium coordinates of the nuclei in
the ground state and the excited state responsible for emission,
respectively. Note that the Franck-Condon integrals cancel out in
this approximation. The justification for the approximations introduced in Eqs. (12) and (13) stems from the fact that for absorption
it is mostly the functions $\varphi_i{}^g$ with small values for i which count,
and they are finite in the vicinity of $Q_o{}^g$, while for emission it is
mostly the functions $\omega_j{}^k$ with small values of j which count, and
they are finite in the vicinity of $Q_o{}^k$. Thus g_{abs} is characteristic
of the participating electronic wave functions at the equilibrium
position of the nuclei in the ground state, while g_{em} is characteristic of the participating electronic wave functions at the equilibrium position of the nuclei in the excited state involved in the
emission process.

The first approximation for g_{abs} and g_{em}, Eqs. (12) and (13),
would predict that these anisotropy factors should assume a constant
value across a single electronic absorption band [20] or across an
emission band. However, as was pointed out by Moffitt and Moscowitz
[13], the first approximation may be inadequate in much the same way
that it may be inadequate for the phenomenon of linear polarization.
This may be the reason for the variation of g_{em} across the emission
band of 1,N^6-ethenoadenine-nicotinamide-dinucleotide bound to gly-
ceraldehyde-3-phosphate dehydrogenase, the fluorescence of which has
a relatively long lifetime and therefore involves a weak transition
[20a]. Moreover, since the anisotropy factors involve also the
cosine of the angle between the electric and magnetic transition
moments, any effect that variation of Q has on this angle may
contribute additionally to the nonconstancy of the anisotropy factors across an electronic absorption or emission band. This is
especially important if the angle between the electric and magnetic
transition moments is close to 90 deg, in which case the rotatory

power should be looked upon as forbidden, although both the electric and magnetic transitions may be allowed.

CPL measurements offer a very convenient way for the investigation of the optical activity of nonisotropic systems. If rotatory motion is frozen in the time scale of the fluorescence lifetime the angular distribution of the excited molecules will depend on the polarization of the exciting beam. To obtain the polarization of the emitted light the contributions of the various photoselected molecules should be properly weighted [20b]. It is of much interest that by measurement of CPL upon excitation by light of the proper polarization and wavelength, the angle between the emitting electric and magnetic transition dipole moments, as well as the magnitude of the magnetic transition moment, can in principle be evaluated [20b]. These physical parameters cannot be obtained by the study of the CD or CPL of isotropic systems, which yield only the scalar product of the electric and magnetic transition moments.

III.   LINEAR POLARIZATION OF FLUORESCENCE

The subject of linear polarization of fluorescence has many aspects, ranging from the application to fundamental spectroscopic studies of small molecules to the probing of complex biological structures such as membranes. Some important trends in the field as applied to biochemical and biological problems are treated at great length in other chapters of this book. In the present section a few of the facets of this field which have a more general, and somewhat problematic, character will be touched upon.

A.   Why Is the Limiting Polarization Rarely Attained?

When one excites with linearly polarized light a frozen solution of randomly oriented molecules in which the absorption and emission transition moments are collinear, the emitted fluorescence observed at 90 deg to the excitation beam and to the direction of the excitation polarization is expected to be of polarization $p = 1/2$, where p is defined as

$$p = \frac{I_{\parallel} - I_{\perp}}{I_{\parallel} + I_{\perp}} \tag{14}$$

$I_{\parallel}$ and $I_{\perp}$ are the intensities of the emitted light polarized parallel and perpendicular, respectively, to the direction of polarization of the excitation beam. The absorption at the red edge of the absorption spectrum and the emission at the blue edge of the fluorescence spectrum are supposed to involve the same vibronic transition and are therefore expected to have the same polarization [11]. (An implicit assumption underlying this statement is that no "hot bands" are involved in the above transitions.) Notwithstanding, the value of $p = 1/2$ is rarely realized [21]. Except for those cases in which the symmetry of the molecules permits a number of nonparallel equivalent transitions between degenerate states, the reason for the above shortcoming is far from being clearly understood [11]. It is not certain at all that a single cause is responsible for the discrepancy in all cases. One or more of the following may possibly explain some cases in which a low value for p was found.

(i) In the study of polarization spectra, all of the fluorescence light, or an unidentified portion of the fluorescence spectrum, is often collected by broad-band filters. The experimental advantages of this procedure are obvious: high intensities of collected light and simplified instrumentation. The justification for this procedure is, however, dubious; it is based on the intuitive notion that all of the emission band, being usually due to a single electronic transition in condensed phases, has the same polarization. As stressed above, the polarization across the emission band may indeed vary. If it does, the limiting polarization of $p = 1/2$ cannot be expected to be realized when the above experimental procedure is followed. Somewhat unfortunately, the measurement of polarization across emission bands does not seem to be popular.

(ii) In the study of polarization spectra care is usually exercised to prevent depolarization due to rotational diffusion by
performing the experiments in highly viscous media, often by freez-
ing the solution to a glass. Glasses are, however, notorious for
their internal strains and concomitant birefringence. Obviously,
any birefringence in the medium will impair the linear polarization
of both the excitation and fluorescence light transversing it, which
will invariably lower the observed polarization values. Moreover,
the extent of the depolarization due to the birefringence of the
medium may depend on wavelength, which may considerably complicate
the observed polarization spectra. Artifacts of this kind were taken
care of and carefully eliminated by Williams by a rather elaborate
procedure in a study of the polarization of phosphorescence [22].
Hopefully, this serious problem is attended to more often than is
evident from the published literature.

If the measured limiting polarization is indeed affected by
incidental strain and birefringence in the medium, it may be expected to exhibit poor reproducibility. It is of interest that the
"randomization factor" introduced by Albrecht to characterize empirically deviations of measured polarizations from ideal behavior
were actually found to vary appreciably from experiment to experiment done with the same substance [23]. Birefringence in the medium
may also disturb measurements of polarization when the luminescence
light is viewed at 90 deg to the direction of propagation of the
excitation beam and in its plane of polarization. Ideally, no polarization should be detected in the luminescence; however, if the
excitation beam is partially depolarized in the medium because of
birefringence, the luminescence will exhibit some polarization.

(iii) The failure to attain the limiting value expected for
the polarization has sometimes been attributed to the presence of
impurities, dimers, aggregates, etc., which have different absorption and emission spectra than the bulk of the fluorescent solute.
This may be especially serious for excitation at wavelengths for
which the absorption coefficient of the substance studied is low.

A more subtle, and rather interesting, possibility is that in the glassy solvent the fluorescent molecules are frozen in different conformations which may have somewhat different spectral properties [24]. Careful studies of the dependence of the fluorescence spectra on the excitation wavelength may shed light on this aspect of frozen solutions of fluorescent compounds.

(iv) In the case of dyes bound to proteins or nucleic acids, the mode of interaction between the biopolymer and the ligand may change upon electronic excitation of the latter. Evidence for such effects will be presented below. Any change in the direction of the electronic transition dipole moment of the ligand relative to the protein accompanying electronic excitation will of course result in some depolarization of the emitted light. (It is conceivable that a free dye molecule embedded in a cage of viscous solvent may also relate to some extent upon electronic excitation due to changes in interaction with the solvent molecules comprising the cage.) Obviously, if a few ligands are bound per protein molecule, any energy transfer that might occur between them may result in depolarization. This latter effect has been studied in great detail by several authors [25].

(v) Last, but not least, the red edge of the absorption spectrum and the blue edge of the fluorescence spectrum most probably contain contributions from "hot bands" [26], i.e., transitions from thermally excited vibrational or so-called librational degrees of freedom. If this is the case, the transitions at the red edge of the absorption spectrum and at the blue edge of the emission spectrum are not pure transitions and do not involve, on the whole, exactly the same pairs of quantum states. The possible consequences of the contribution of "hot bands" to the spectra as regards the polarizations of these transitions have apparently not been adequately considered.

B.   Long-Range Nonradiative
Transfer of Electronic Excitation Energy

The considerations regarding the polarization of absorption and emission bands are relevant to the phenomenon of long-range transfer of electronic excitation energy by the dipole-dipole mechanism. As usually formulated, the probability of energy transfer per unit time $n_{A \to B}$ from a donor molecule A to an acceptor molecule B is given by the expression

$$n_{A \to B} = 8.8 \times 10^{-25} \eta \kappa^2 n^{-4} \tau^{-1} r^{-6} \int_0^\infty f(\bar{\nu}) \epsilon(\bar{\nu}) \bar{\nu}^{-4} \, d\bar{\nu} \tag{15}$$

where η is the fluorescence quantum yield of the donor in the absence of the acceptor, n is the refraction index of the medium, τ is the fluorescence lifetime of the energy donor, r is the distance between A and B, $f(\bar{\nu})$ is the normalized fluorescence intensity of the donor at wave number $\bar{\nu}$, $\epsilon(\bar{\nu})$ is the extinction coefficient of the acceptor at $\bar{\nu}$, and κ is a geometric factor determined by orientation in space of the transition dipole moments of donor and acceptor. In the derivation of Eq. (15) it is assumed that the emission band of the donor and the absorption band of the acceptor are characterized each by a single transition dipole moment vector of constant direction across the bands. κ then assumes the form

$$\kappa = \cos \alpha_{AB} - 3 \cos \alpha_A \cos \alpha_B \tag{16}$$

where α_{AB} is the angle between the transition dipole moments of A and B, and α_A and α_B are the angles between the transition dipole moment of A and B, respectively, and the line joining A and B. The above assumption regarding the polarization of the absorption and emission transitions may not be justified: under some circumstances Eqs. (15) and (16) will then have to be modified.

In the most general case, the absorption or emission transitions may be described by three orthogonal incoherent transition dipole moments which may vary with wavelength. Let us denote these transition dipole moments as $\underline{x}_1(\bar{\nu})$, $\underline{x}_2(\bar{\nu})$, and $\underline{x}_3(\bar{\nu})$ for the emission of the donor and $\underline{y}_1(\bar{\nu})$, $\underline{y}_2(\bar{\nu})$, and $\underline{y}_3(\bar{\nu})$ for the absorption of the acceptor. Define κ_{ij} as

$$\kappa_{ij}(\bar{\nu}) = \cos \alpha_{x_i y_j} - 3 \cos \alpha_{x_i} \cos \alpha_{y_j} \; ; \quad i = 1, 2, 3, \quad j = 1, 2, 3 \tag{17}$$

where $\alpha_{x_i y_j}$ is the angle between $\underline{x}_i$ and $\underline{y}_j$, and α_{x_i} and α_{y_j} are the angles between $\underline{x}_i$ and $\underline{y}_j$, respectively, and the line joining A and B. The generalized equation for the probability of transfer of energy from donor to acceptor will then assume the form

$$n_{A \to B} = 8.8 \times 10^{-25} \; \eta n^{-4} \tau^{-1} r^{-6} X^{-2} Y^{-2} \int [\sum_{i=1}^{3} \sum_{j=1}^{3} x_i^2 y_j^2 \kappa_{ij}^2(\bar{\nu})] f(\bar{\nu}) \epsilon(\bar{\nu}) \bar{\nu}^{-4} \, d\bar{\nu} \tag{18}$$

where

$$X^2 = \sum_{i=1}^{3} x_i^2 \quad \text{and} \quad Y^2 = \sum_{j=1}^{3} y_j^2$$

No systematic experimental test of the variation of the probability of energy transfer with the orientation of the donor and of the acceptor has as yet been devised. It is obvious from Eq. (18) that any such test, when rendered possible, will require knowledge on the polarization properties of the emission band of the donor and of the relevant absorption band of the acceptor.

Equation (18) may have some practical significance. In the application of long-range nonradiative energy transfer to the study of distances between chromophores in biochemical systems, such as proteins and nucleic acids, prior knowledge of the directions of the transition moments of the donor and acceptor is required and is usually missing [27]. It can be shown from Eqs. (17) and (18) that if $\underline{x}_1$, $\underline{x}_2$, $\underline{x}_3$, $\underline{y}_1$, $\underline{y}_2$, and $\underline{y}_3$ are all finite and/or dependent on wave number, the dependence of $n_{A \to B}$ on the orientations of the donor and the acceptor is less pronounced than when the transition moments are single vectors whose directions do not vary with wave number. In the idealized extreme case in which $\underline{x}_1 = \underline{x}_2 = \underline{x}_3$ and $\underline{y}_1 = \underline{y}_2 = \underline{y}_3$, $n_{A \to B}$ does not depend on orientation at all. (The author is very grateful to L. Stryer for fruitful and interesting discussions of this subject.) It may be noted that cases of the kind $\underline{x}_1 = \underline{x}_2$, so-called planar oscillators, are realized in practice for molecules of appropriate symmetry properties [21]. Proper choice of chromophores which fulfill these requirements to a greater or lesser extent may help to alleviate the orientational problem encountered in the use of nonradiative energy transfer as a spectroscopic ruler [27]. It is pertinent to note that in case the emission band of the donor overlaps more than one electronic band of the acceptor, κ_{ij} is likely to show marked dependence on wave number. One may therefore avoid the extreme values that the orientational factor can assume by having the fluorescence spectrum of the donor overlap more than one absorption band of the acceptor, and by choosing chromophores which have mixed polarizations in the relevant absorption and emission bands.

C. The Red-Edge Effect

Weber and Shinitzky [28] have described an interesting and intriguing phenomenon regarding the failure of concentration depolarization of fluorescence when the excitation is performed by light absorbed at the red edge of the absorption spectrum. It is quite a general phenomenon that when substances in highly viscous media are excited at the bulk of the absorption bands, the polarization of their fluorescence steadily decreases with increase of their concentration. This behavior is expected, since the absorption and emission spectra of most materials show some overlap, thus allowing long-range nonradiative energy transfer to take place between neighboring molecules when the average intermolecular distance is favorable. Since the solute molecules are not aligned parallel to each other, such transfer of energy will tend to randomize the direction of the emitting molecules and cause depolarization. However, this process does not take place when the molecules are excited at the red edge of their absorption spectrum. Why?

Apparently no clear-cut explanation of this phenomenon has been convincingly put forward as yet. The red-edge effect may reflect some shortcoming of the theory of energy transfer or an erroneous interpretation of the experimental results. It seems to be of interest to explore some possible causes for this disconcerting effect.

As to the theoretical aspects of the problem, it may be recalled that a controversy exists whether the theory of long-range nonradiative energy transfer, as formulated by Förster, holds also for transfer of energy among molecules of the same kind. Robinson and Frosch [29] claim that in the latter case the interacting molecules oscillate coherently, and the so-called very weak coupling treatment of Förster does not hold. On the other hand, Förster [30] and Bennett and Kellogg [31] argue that the perturbations induced by vibrations and disturbances of the environment ruin the coherence of the interaction between like, as well as unlike, molecules. The trend has been to accept the latter arguments. However, is it possible that these arguments fail in the case of like molecules embedded in a frozen medium at relatively low temperatures and excited at a frequency where both absorption and emission occur?

A few reservations have been raised as to the interpretation of the experiments. These reservations have in common the notion that upon excitation at the red edge of the absorption spectrum one tends to excite an entity which is different from the bulk of the dissolved material and which has an emission spectrum which poorly overlaps the absorption spectrum of the bulk solute. This entity can be impurity molecules, dimers or aggregates of the solute, or solute molecules frozen in strained conformations [24]. Still another possibility exists that excitation at different portions of the absorption spectrum results in different emission spectra due to incomplete thermal equilibration of the molecules in the excited state. A

shift of the fluorescence spectrum to the red will result in a reduced overlap with the absorption spectrum and may account for the
reduced probability of energy transfer. Recent studies using time
resolved spectroscopy show indeed that the time needed for thermal
equilibration may not be negligibly short compared to the lifetime
of the excited state [32,32a]. Such effects may be specially pronounced at the low temperatures and high solvent viscosities at
which polarization studies are usually conducted; the relevant absorption and fluorescence spectra should thus be investigated under
the same conditions at which the depolarization experiments are
performed. In this connection it is relevant to mention recent
findings that the emission spectrum of 1-anilinonaphthalene-8-sul-
fonate bound to bovine serum albumin is appreciably shifted to longer
wavelengths when excited at the red edge of the absorption spectrum,
the shift being larger the lower the temperature [33]. This shift
could possibly account, at least in part, for the red-edge effect
observed in this system by Anderson and Weber [34]. Shifts in the
emission spectra of several dyes upon excitation at the red edge of
the absorption spectrum were observed by Weber and Shinitzky [28]
but were considered insufficient to explain the observed lack of
concentration depolarization.

IV. CIRCULAR POLARIZATION OF LUMINESCENCE

As explained above, circular polarization of luminescence, CPL,
is related to the chirality of the emitting molecules in the electronically excited state in the same way that circular dichroism,
CD, is related to the chirality of the absorbing molecules in the
ground state. If the conformation of the molecules is the same in
the ground and excited states, then to a first approximation the
anisotropy factors for absorption and emission, g_{abs} and g_{em},

respectively, will assume the same value [cf. Eqs. (12) and (13)].
This is not often the case. CPL thus yields information not attain-
able otherwise. Furthermore, CPL has some characteristic features
which, if properly recognized, may be exploited to shed light on a
variety of problems. It is selective, since it applies to luminescent chromophores only. This property may be of tremendous advantage in biochemical and biological systems. It can yield information
about the optical rotatory power of transitions which are too weak
to be seen in absorption, such as the triplet-singlet transitions involved in phosphorescence, or otherwise nearly-forbidden transitions.
Other potentialities of the method may be illustrated by a few
studies on CPL to be described in the following section.

A. A Few Notes Regarding Instrumentation

A sensitive instrument capable of measuring low levels of circular polarization in luminescence light has been recently described
[35]. Since no such commercial instrument is available, a few general comments regarding the instrumental aspects of the study of CPL
may not be out of place.

(1) The extent of circular polarization of the light emitted
by chiral molecules is expected to fall in most cases in the range
of 10^{-2}-10^{-4}. This is the typical range of values found for g_{abs},
and g_{em} is likely to be of similar order of magnitude, although different values may of course be sometimes encountered [36,36a]. A
modulation technique in which the circularly polarized component is
modulated in time while the unpolarized light is unaffected, coupled
with phase-sensitive amplification of the electronic signal produced
by the modulated light, was found to be effective under these circumstances [35].

(2) To obtain a reasonable signal-to-noise ratio, e.g., better
than 10, the number of photons collected per measurement should exceed 10^{10} for $g_{em} = 10^{-4}$, assuming that the main source of noise is

due to fluctuations in photon counting [35]. Thus, light intensity
is an aspect of such an instrument which should not be overlooked.

   (3) Artifacts due to linear polarization should be carefully
avoided. This is especially important since the extent of circular
polarization of the fluorescent light is usually low, while linear
polarization may be large if rotational diffusion of the fluorescent
molecules is not fast enough. Depolarization of the excitation light,
observation of the fluorescence at zero degrees to the excitation
beam, in addition to the phase-sensitive amplification, were found
to be effective in overcoming this problem. Obviously, stray light
should be carefully avoided. This was achieved by use of monochro-
mators for both the excitation and emission light and suitable aux-
iliary optical filters [35,35a,35b].

   (4) A means of calibration of the instrument is essential. A
simple accessory which produces at will known mixtures of unpolarized
and circularly polarized light has been described and was found to be
extremely useful [35].

B. CPL as a Source of Information
about the Emitting Molecule in the Excited State

   As discussed in Section II,B, the CPL of a molecule depends upon
its conformation in the excited state. The information that is ex-
tractable from CPL in terms of conformational parameters will depend,
of course, on the complexity of the system studied; it may range from
detailed molecular data in favorable cases to qualitative information
in complicated systems. An example of a case of the former type will
be presented.

   1,1'-Bianthracene-2,2'-dicarboxylic acid cannot rotate freely
around the single bond which connects the two anthracene rings in
the molecule, owing to steric hindrance. It can therefore be resolved into two optical isomers. The emission anisotropy factor was
measured across the fluorescence spectrum of each of the enantiomers

of this compound and was found to be two- to threefold smaller in absolute magnitude than its absorption anisotropy factor at the long wavelength absorption band [37]. This difference between g_{abs} and g_{em} obviously reflects a change in the molecular conformation of this substance upon electronic excitation, such as a change in the di-hedral angle between the planes of the two anthracene rings. A quantitative calculation of this change in conformation by a quantum-mechanical extension of the consistent force field method was under-taken by Schlessinger and Warshel [38]. The values to be expected for g_{abs} and g_{em} for this compound were calculated for various di-hedral angles between the two anthracene rings, and the results were compared with the experimental data. It was thus deduced that this angle changes by about 20 deg upon electronic excitation. Hopefully, more molecules will lend themselves to such treatment for the study of their conformation in the excited state.

While a detailed analysis of CPL data may be impractical for complicated systems, it may still be useful as a very sensitive diag-nostic tool for changes in conformation upon electronic excitation. As discussed in Section II,B, if the molecular conformation and en-vironment are the same in the ground and excited states, g_{abs} at the red edge of the absorption spectrum and g_{em} are expected to be of equal sign and magnitude. This was found to be approximately the case, for example, for the tryptophan residues of Staphylococcal nuclease [38a] and azurin [38b], and for the pyridoxal cofactor in phosphorylase b [38c]. In contrast, a discrepancy between g_{abs} and g_{em} is indicative of a difference in the conformation of the mole-cule in the ground and excited states. Numerous cases of interest of this kind were observed.

An interesting study on the difference between g_{abs} and g_{em} was performed by Emeis and Oosterhoff [36,39] on trans-β-hydrindanone. Both g_{abs} at the last absorption band and g_{em} of this compound are exceedingly large, being of the order of 0.15 and 0.03, respectively (the anisotropy factors are not constant across the electronic bands in this compound). The large difference between g_{abs} and g_{em} was rationalized in terms of the geometry of the carbonyl carbon in the ground and excited states, this carbon and its substituents changing from planar to pyramidal geometrical arrangements upon electronic excitation. The molecule has additional electric dipole strength, and hence additional total intensity, when in the pyramidal arrangement, but this additional electric transition dipole does not add to the rotational strength of the molecule since it is orthogonal to the magnetic transition dipole moment. The anisotropy factor is thus expected to be smaller when the carbonyl chromophore is in the pyramidal configuration [39]. From a similar study of the CD and CPL of the α-diketone d-camphorquinone, Luk and Richardson [39a] concluded that the chirality of the dicarbonyl group is the same in the ground and excited states but that the dihedral twist angle is smaller in the fluorescent state than in the ground state.

Differences between g_{abs} and g_{em} do not always lend themselves to neat interpretation in basic terms, but they may still be very instructive empirically. A typical example of this kind is the complex of chymotrypsin with 2-p-toluidinyl naphthalene-6-sulfonate, which binds to the protein at a specific site (which is not the active site). In this case the dye assumes chiral properties owing to its interaction with the asymmetric binding site on the surface of the protein. A dramatic difference was found between g_{abs} and g_{em} of the bound dye, the latter being about tenfold smaller than the former [40]. The mode of interaction between the protein and the ligand must thus be different when the dye is in the ground state than when in the excited state. Caution must therefore be exercised in drawing conclusions concerning the detailed characteristics of the binding of dye and protein in the ground state from fluorescence data. Furthermore, changes in the mode of interaction between the protein and the ligand upon electronic excitation may involve a change in the direction of the electric transition dipole moment of the ligand relative to the protein, which may account for the lower values of the limiting polarization sometimes found for chromophores bound to proteins compared to those found for the free chromophores in frozen solution. If this change in relative orientation of protein and ligands occurs in the time scale of the fluorescence decay and its rate is temperature-dependent, it may complicate the interpretation of fluorescence depolarization in terms of the rotational diffusion of the macromolecule. A similar, though less pronounced, difference between g_{abs} and g_{em} was also found for the anthraniloyl chromophore covalently bound at the active site of chymotrypsin.

In the study of the CD and CPL of various cyclic dipeptides containing aromatic side-chains it was found that the optical rotatory power characteristic of the ground state vanishes upon electronic excitation in fluid media [41]. However, essentially the same rotatory power was found for the ground and excited states in highly viscous media. A change in molecular conformation thus occurs upon excitation in fluid media essentially before light emission occurs, but the molecular conformation can be frozen by a high viscosity of the medium.

A case of special interest in which the optical activity in the ground state is different from that in the excited state is that of chlorophyll dimers. Chlorophyll a is known to dimerize in carbon tetrachloride solution. The absorption bands in the visible absorption spectrum of the dimer show well-defined circular dichroism; however, no circular polarization could be detected in its fluorescence band when dissolved in a fluid solvent [42,42a]. It may be noted that similarities between the structures of the dimers in solution and chlorophyll aggregates in chloroplasts were inferred from circular dichroism data [43]. Since it is the excited state which is of prime interest in the case of chlorophyll, one should thus exercise extreme caution in taking the dimers as models for chlorophyll aggregates in vivo.

C. Behavior of the
Anisotropy Factor across the Emission Band

Since fluorescence in condensed media usually involves transitions between a single pair of electronic levels, g_{em} should be, to a first approximation, constant across the emission band [see Eq. (13)]. This was found to be the case, for example, for Staphylococcal nuclease [38a] and azurin [38b], the molecules of which contain a single tryptophan residue. As was pointed out above, the first approximation may not be satisfactory; thus g_{em} does not <u>have</u> to assume a fixed value. Notwithstanding, g_{em} is expected to show simpler dependence on change of wavelength than g_{abs}, since in the absorption spectra different electronic transitions express themselves, for which g_{abs} may assume entirely different values even within the limits of the first approximation. It is thus advisable to be on guard if g_{em} is found to show erratic changes across the emission spectrum. Such behavior was indeed found in a few cases and proved to be, upon additional examination, very informative. Two examples will be cited below.

Acridine dyes have been known for a long time to acquire asymmetric properties when bound to helical polyglutamic acid [44]. Complexes of acridine dyes with this polymer were indeed found to exhibit some circular polarization in their fluorescence [38a,45]. Opposite sign but equal absolute magnitude was found for g_{em} of such complexes with helical poly-L- or poly-D-glutamic acid, the circular polarization disappearing under conditions at which the polymer assumes the random-coil conformation. The emission anisotropy factors were found, however, to vary markedly across the emission bands. Thus, for the complex of poly-L-glutamic acid with 9-aminoacridine, g_{em} assumed the values of $+2.5 \times 10^{-4}$ and of -16×10^{-4} at the short wavelength region (450 nm) and at the long wavelength region (550 nm),

respectively, of the emission band. Also for the complex of poly-A
with 9-aminoacridine, g_{em} varied markedly across the emission band
[35a]. Study of the emission spectrum at various dye-to-polymer
ratios revealed that the fluorescence spectrum is a superposition of
the emission from monomer and dimer (or aggregated dye molecules) at
the short and long wavelength regions of the spectrum, respectively.
Thus the anisotropy factors for the monomeric and dimeric dye molecules in the excited state are available. Obviously, any explanation
as to the structure of the dimers should also account quantitatively
for their optical rotatory power.

Another example in which the emission anisotropy factor was
found to vary markedly with wavelength was the complex of bovine
serum albumin with 1-anilinonaphthalene-8-sulfonate [45]. Approximately similar values for g_{em} were found for various dye-to-protein
ratios in the range of 1 to 5. This is in line with previous sugges-
tions that the first five sites for the dye on bovine serum albumin
have roughly the same binding constants; these sites are thus roughly
equally populated when the dye is added to the protein. (The CD
spectrum was found to depend on the dye-to-protein·ratio [46]; the
difference between the behavior of the CD and the CPL may be due to
a number of reasons.) One simple explanation for the variation of
g_{em} with wavelength is that the dye molecules at the various sites
emit at somewhat different wavelengths and have different optical
activity. The optical activity measured across the composite emission spectrum cannot thus be expected to be constant. However,
another factor which is of much interest may be operative as well.
As quoted above, the emission spectrum of 1-anilinonaphthalene-8-
sulfonate bound to bovine serum albumin depends on the excitation
wavelength, the dependence being more pronounced at low temperatures.
The thermal relaxation of the dye molecule in the excited state thus
seems to proceed in the time scale of the lifetime of the excited
state. The overall emission spectrum observed statically is therefore a superposition of spectra emitted at various time intervals
following the moment of excitation. During its thermal relaxation

in the excited state, the molecule most probably changes its interaction with its environment. Its induced optical activity may therefore change with time and hence with its momentary emission spectrum.
At this stage of our knowledge, this explanation is somewhat specu-
lative, but it may be interesting to put it to more extensive experimental test.

D. CPL as a Diagnostic Tool for the Study of Macromolecular Conformation

     Similarly to CD, the CPL of chromophores in, or attached to,
macromolecules usually stems from asymmetric perturbations induced
by the environment of the chromophores. CPL can thus substitute for
CD in many cases in the study of macromolecules by chromophoric
probes if the chromophores of interest are fluorescent. CPL may even
sometimes be preferable to CD if the optical activity of the system
in the excited state happens to be larger than its optical activity
in the ground state. Furthermore, more chromophores contribute to
CD than to CPL; changes in conformation may therefore produce larger
changes in CPL than in CD because cancellation of signals of opposite
sign coming from different chromophores is more likely to occur when
more chromophores are involved. A few examples will be presented
below.

     The CPL of the tryptophan and tyrosine emission of fully re-
duced and denatured proteins was invariably found to be zero within
experimental error [38a]; the CPL found for native proteins thus reflects the effect of the secondary and tertiary structure of the
proteins on the fluorescent side chains. In the case of subtilisin
Novo and subtilisin Carlsberg a change of pH from 5.0 to 8.3 was
found to affect the CPL of the tyrosine residues in both proteins
and the CPL of the tryptophan residues in the Novo protein (the
contributions from the tyrosine and the tryptophan residues were
readily resolved). Conformational changes in the molecules may thus

be, at least in part, the cause for the pH dependence of their enzymatic activity [35b]. CPL was applied to the study of subunit
interactions in an antibody molecule and in its Fab' and Fv fragments
[47]. It was shown that the separate light and heavy chains have
different conformations than the intact parent molecules, at least
at the environment of the tryptophan residues. The resistance of
the molecules to denaturation by urea increases with increasing
length of the constituent chains, and different regions of these
molecules unfold with varying ease. At least in the case of Fv, the
unfolding proceeds through intermediate structures which possess a
CPL spectrum that is markedly different from that of the native conformation [47]. Changes were observed in the CPL of antibodies upon
binding of the corresponding antigens. Changes in CPL were also
observed in their Fab' fragments upon binding of the corresponding
antigens, but these were markedly different from those observed for
the whole antibody molecules, thus implying a change in conformation
of the Fc fragment upon binding of antigen to antibody [48]. The
disulfide bonds at the hinge region of the antibody were shown to
be involved in the transmission of the conformational change from
the Fab to the Fc [49].

CPL has also been successfully applied to the study of interaction reactions of proteins. The interaction between bovine pan-
creatic trypsin inhibitor with trypsin and chymotrypsin was investigated by study of the CPL of an anthraniloyl chromophore selectively
attached to lysine 15 of the inhibitor [50]. Progressive structural
changes were found to occur at the adenine subsites in the tetra-
meric molecules of glyceraldehyde-3-phosphate dehydrogenase upon
cumulative binding of $1,N^6$-ethenoadenine-nicotinamide-dinucleotide.
The greatest structural change in the protein occurs upon binding
of the first molecule of coenzyme per tetramer [51]. These results
offer a molecular explanation for the negative cooperativity in the
binding of the coenzyme. By use of CPL as a probe for the binding

sites of various dehydrogenases it was indicated that those dehydrogenases which bind their coenzyme noncooperatively have similar,
if not identical, adenine subsites, whereas in glyceraldehyde-3-
phosphate dehydrogenase, a strongly cooperative enzyme, the adenine
subsite has a different structure [20a].

Some systems are optically active when in the excited state but
are optically inactive (within experimental error) when in the ground
state. CPL may thus be used as a diagnostic tool where CD is inap-
plicable. The complex between anti-dansyl antibody and the dansyl
hapten is such an example. By use of CPL it could thus be shown
that the dansyl hapten is embedded in an asymmetric environment when
bound to anti-dansyl antibody. On the other hand, when 1-alanyl-2-
dansyl diaminoethane binds to anti-alanyl antibody, neither CD nor
CPL could be detected in the dansyl spectra. In this case the dansyl
group is apparently outside the field of influence of the protein
[17a,52].

The CPL of the emission of terbium ions bound at the iron binding sites of human serum transferrin revealed that the two sites per
molecule are equivalent, and that they are similar in structure to
the iron binding sites of avian egg conalbumin [36a]. It should be
noted that the CD of the terbium absorption could not be studied
because of the very low absorption intensity.

In common with other experimental methods, CPL as a tool for
the study of the optical rotatory power of molecules has some characteristic limitations. It is applicable only to luminescent molecules, and then only to the electronic transitions which are involved
in the luminescence process (usually $S_1 \rightarrow S_0$ and $T_1 \rightarrow S_0$). These limitations may turn out, however, to be a blessing, especially in the
study of complex systems, where it often pays to trade generality
for specificity. Some of the above examples may serve as an illustration of this point.

V. CONCLUSION

Fluorescence, in general, and fluorescence polarization, in particular, are well-established and active fields of research. The accomplishments in these fields are numerous indeed. It is therefore surprising, and somewhat disconcerting, that a few fundamental points have not been fully appreciated or are not yet well understood. It is, of course, important to be able to live with problems, but one should at least be aware of their existence.

This chapter has possibly disappointed some readers who hoped fluorescence polarization to be a simple, unequivocal, and straightforward experimental technique. It is hoped, however, that few readers were wrongly impressed that it is a dull subject or a field of little consequence. The opposite is no doubt true; this field of research is of tremendous interest and importance in its implications about the nature of matter and in the possibilities it opens up as a tool in chemistry and biology. Not least of all, it offers ample scope for imagination and play.

REFERENCES

1.   P. Selényi, Ann. Physik, 35, 444 (1911).

2.   S. Freed and S. I. Weisman, Phys. Rev., 60, 440 (1941).

3.   H. Kuhn, D. Möbius, and H. Bücher, in Techniques of Chemistry (A. Weissberger and B. W. Rossiter, eds.), Vol. 1, Part III B, Wiley-Interscience, New York, 1972, pp. 577-702.

4.   G. Weber, Advan. Protein Chem., 8, 415 (1953).

5.   G. Weber, J. Chem. Phys., 55, 2399 (1971).

6.   L. Stryer, Science, 162, 526 (1968).

7.   G. Weber, Trans. Faraday Soc., 50, 552 (1954).

8.   G. Herzberg and E. Teller, Z. Physik. Chem., B21, 410 (1933).

9. H. Sponer and E. Teller, Rev. Mod. Phys., 13, 75 (1941).

10. A. C. Albrecht, J. Chem. Phys., 33, 156 (1960).

11. A. C. Albrecht, J. Mol. Spectrosc., 6, 84 (1961).

12. A. C. Albrecht and W. T. Simpson, J. Chem. Phys., 23, 1480 (1955).

13. W. Moffitt and A. Moscowitz, J. Chem. Phys., 30, 648 (1959).

14. W. Liptay, in Modern Quantum Chemistry (O. Sinanoglu, ed.), Part III, Academic Press, New York, 1965, pp. 81-92.

15. L. Strackee, Photochem. Photobiol., 15, 253 (1972).

16. A. N. Sevchenko, G. P. Gurinovich, and K. N. Solev'ev, Sov. Phys. Dokl., 5, 808 (1961).

17. M. Gouterman and L. Stryer, J. Chem. Phys., 37, 2260 (1962).

17a. J. Schlessinger, I. Z. Steinberg, and I. Pecht, J. Mol. Biol., 87, 725 (1974).

17b. J. Schlessinger and I. Z. Steinberg, in preparation.

18. W. Kuhn, Ann. Rev. Phys. Chem., 9, 417 (1958).

19. E. U. Condon, W. Altar, and H. Eyring, J. Chem. Phys., 5, 753 (1937).

20. A. Moscowitz, in Modern Quantum Chemistry (O. Sinanoglu, ed.), Part III, Academic Press, New York, 1965, pp. 31-44.

20a. J. Schlessinger, I. Z. Steinberg, and A. Levitzki, J. Mol. Biol., 91, 523(1975).

20b. I. Z. Steinberg and B. Ehrenberg, J. Chem. Phys., 61, 3382 (1974).

21. P. P. Feofilov, The Physical Basis of Polarized Emission, English translation, Consultants Bureau, New York, 1961.

22. R. Williams, J. Chem. Phys., 30, 233 (1959).

23. A. C. Albrecht, J. Amer. Chem. Soc., 82, 3813 (1960).

24. R. M. Hochstrasser, in Probes of Structure and Function of Macromolecules and Membranes (B. Chance, C. Lee, and J. K. Blasie, eds.), Academic Press, New York, 1971, pp. 57-64.

25.   I. Z. Steinberg, Ann. Rev. Biochem., 40, 83 (1971).

26.   L. E. Erickson, J. Luminescence, 5, 1 (1972).

27.   L. Stryer and R. P. Haugland, Proc. Natl. Acad. Sci. USA, 58, 719 (1967).

28.   G. Weber and M. Shinitzky, Proc. Natl. Acad. Sci. USA, 65, 823 (1970).

29.   G. W. Robinson and R. P. Frosch, J. Chem. Phys., 38, 1187 (1963).

30.   T. Förster, in Modern Quantum Chemistry (O. Sinanoglu, ed.), Part III, Academic Press, New York, 1965, pp. 93-137.

31.   R. G. Bennett and R. E. Kellogg, Progr. Reaction Kinetics, 4, 215 (1966).

32.   L. Brand and J. R. Gohlke, Ann. Rev. Biochem., 41, 843 (1972).

32a. A. Grinvald and I. Z. Steinberg, Biochemistry, 13, 5170 (1974).

33.   J. Schlessinger and I. Z. Steinberg, in preparation.

34.   S. R. Anderson and G. Weber, Biochemistry, 8, 371 (1969).

35.   I. Z. Steinberg and A. Gafni, Rev. Sci. Instr., 43, 409 (1972).

35a. A. Gafni, J. Schlessinger, and I. Z. Steinberg, Israel J. Chem., 11, 423 (1973).

35b. J. Schlessinger, R. S. Roche, and I. Z. Steinberg, Biochemistry, 14, 255 (1975).

36.   C. A. Emeis and L. J. Oosterhoff, Chem. Phys. Lett., 1, 129 (1967).

36a. A. Gafni and I. Z. Steinberg, Biochemistry, 13, 800 (1974).

37.   A. Gafni and I. Z. Steinberg, Photochem. Photobiol., 15, 93 (1972).

38.   J. Schlessinger and A. Warshel, Chem. Phys. Lett., 28, 380 (1974).

38a. I. Z. Steinberg, J. Schlessinger, and A. Gafni, in Peptides, Polypeptides and Proteins (E. R. Blout, F. A. Bovey, M. Goodman, and N. Lotan, eds.), Wiley, New York, 1974, pp. 351-369.

38b. A. Grinvald, J. Schlessinger, I. Pecht, and I. Z. Steinberg, Biochemistry, in press.

38c. S. Veinberg, S. Shaltiel, and I. Z. Steinberg, Israel J. Chem., 12, 421 (1974).

39. C. A. Emeis and L. J. Oosterhoff, J. Chem. Phys., 54, 4809 (1971).

39a. C. K. Luk and F. S. Richardson, J. Amer. Chem. Soc., 96, 2006 (1974).

40. J. Schlessinger and I. Z. Steinberg, Proc. Natl. Acad. Sci. USA, 69, 769 (1972).

41. J. Schlessinger, A. Gafni, and I. Z. Steinberg, J. Amer. Chem. Soc., 96, 7396 (1974).

42. A. Gafni, H. Hardt, and I. Z. Steinberg, in Abstracts of the 42nd Meeting of the Israel Chemical Society, p. 109, Dec., 1972.

42a. A. Gafni, H. Hardt, J. Schlessinger, and I. Z. Steinberg, Biochim. Biophys. Acta, in press.

43. E. A. Dratz, A. J. Schultz, and K. Sauer, in Energy Conversion by the Photosynthetic Apparatus, pp. 93-96, Brookhaven Symposia in Biology, No. 19, 1972.

44. E. R. Blout and L. Stryer, Proc. Natl. Acad. Sci. USA, 45, 1591 (1959).

45. I. Z. Steinberg, J. Schlessinger, and A. Gafni, in Abstracts of the 8th FEBS Meeting, Amsterdam, Aug., 1972.

46. S. R. Anderson, Biochemistry, 8, 4838 (1969).

47. J. Schlessinger, I. Z. Steinberg, D. Givol, and J. Hochman, FEBS Lett., 52, 231 (1975).

48. D. Givol, I. Pecht, J. Hochman, J. Schlessinger, and I. Z. Steinberg, Progress in Immunology II, Vol. 1 (Brent and Holbrow, eds.), North Holland, 1974, pp. 39-48.

49. J. Schlessinger, I. Z. Steinberg, D. Givol, J. Hochman, and I. Pecht, submitted.

50. Y. Elkana, in Proteinase Inhibitors, Bayer Symp. V (H. Fritz, H. Tschesche, L. J. Greene, and E. Truscheit, eds.), Springer-Verlag, Berlin, 1974, p. 445.

51. J. Schlessinger and A. Levitzki, J. Mol. Biol., 82, 547 (1974).

52. J. Schlessinger, I. Pecht, and I. Z. Steinberg, in Abstracts of the 42nd Meeting of the Israeli Chemical Society, p. 111, Dec., 1972.

Chapter 4

POLARIZED EXCITATION ENERGY TRANSFER

Robert E. Dale* and Josef Eisinger
Bell Laboratories
Murray Hill, New Jersey

II.   THEORETICAL BACKGROUND 117
      A. Energy Transfer Theory 117
      B. Fluorescence Depolarization 123
      C. Transfer Depolarization and the Orientation Factor . . 125
III. SELECTED MODELS 134
IV.   EXPERIMENTAL METHODS 228
      A. Depolarization Factors 228
      B. Transfer Depolarization Factors 235
      C. Transfer Efficiency 237
V.   ILLUSTRATIONS OF DISTANCE DETERMINATIONS 239
      A. The Structure of Transfer RNA 241
      B. Distribution of Donor-Acceptor Separations 245
VI.   CONCLUDING REMARKS 245

      APPENDIX A: POLARIZED EMISSION SPECTROSCOPY 247
      APPENDIX B: THE ORIENTATION FACTOR AND TRANSFER
                   DEPOLARIZATION — ILLUSTRATIVE SELECTED
                   MODELS 257
      APPENDIX C: LUMINESCENCE DEPOLARIZATION BY RAPID,
                   RESTRICTED MOTION 271

      ADDENDUM 279

      REFERENCES 280

* Present address: Mergenthaler Laboratory for Biology and
McCollum-Pratt Institute, The Johns Hopkins University,
Baltimore, Maryland

I. INTRODUCTION

An isolated luminophore in an excited electronic state may
lose its excitation energy either by emitting radiation or by nonradiative de-excitation, the rates of these processes depending on
the nature of the luminophore and its environment. In the presence
of other chromophores, it may also donate its excitation nonradia-
tively. Förster [1,2] has developed a theory for calculating the
rate of such a transfer process in the very weak, dipole-dipole
coupling limit which is generally applicable to dyes and other
chromophores of biochemical interest in solution and thereby opened
the way for using the measurement of energy transfer efficiencies
as a means of determining intramolecular distances in macromolecules.
The last few years have seen an increasing number of applications of
this technique to various problems in molecular biology [3-8].

The energy transfer efficiency predicted by the Förster theory
is appreciable over distances greater than a few angstroms if each
of the following conditions is met: (1) The donor quantum yield
must be sufficiently high, of the order of 10% or more. (2) In
order that energy be conserved in the transfer, there must exist a
reasonable overlap between the donor emission spectrum and the
acceptor absorption spectrum. (3) The transition dipoles of the
donor and acceptor chromophores must be suitably oriented with respect to each other. (4) The donor acceptor separation (R) must not
be too great, the limiting value depending on how well the other
conditions are met. For shorter distances, so-called strong interactions (such as charge transfer and exchange) become predominant.

As will be seen below, these conditions are sufficiently well-
satisfied in many experimental situations to have encouraged several
workers to develop energy transfer as a "spectroscopic ruler" [9].
However, a serious limitation to the use of this technique for determining intramolecular distances is presented by condition (3).
The transfer rate is proportional to an orientation factor (κ^2), and,
except for x-ray crystallography, there exist no direct experimental

methods for finding its value. Even where such crystallographic results are available, it is not clear that the equilibrium orientations
of the chromophores will be unchanged in solution, and, in either
case, the problem of averaging the orientation factor over any relative motion of these groups would remain. The present paper suggests
how polarized fluorescence spectroscopy and, more particularly, the
measurement of depolarization during transfer can be used to estimate
values of κ^2 for a broad range of models descriptive of the average
orientation and motional freedom enjoyed by donors and acceptors.
While this approach does not generally permit the precise determination of the orientation factor, it often reduces the limits of uncertainty to an acceptable level.

II.   THEORETICAL BACKGROUND

A.   Energy Transfer Theory

1. Transfer Rates

In the absence of strong interactions, including interactions
which require a spatial overlap between the electronic wave function
of the donor and acceptor to compete with other de-excitation modes
[10], the one-way transfer rate from an excited donor (D) to a suitable acceptor (A) is given by

$$k_T = \frac{9(\ln 10)\Phi_D \kappa^2}{128\pi^5 N' n^4 \tau_D R^6} \, J(\tilde{\nu}) = \frac{1}{\tau_D}\left(\frac{R_o}{R}\right)^6 \tag{1}$$

N' is Avogadro's number per cubic centimeter, R the donor-acceptor
separation, Φ_D and τ_D the donor emission quantum yield and lifetime,
respectively, in the absence of the acceptor, $J(\tilde{\nu})$ the spectral overlap integral of donor emission and acceptor absorption:

$$J(\tilde{\nu}) = \int_0^\infty \epsilon(\tilde{\nu})f(\tilde{\nu})\,\frac{d\tilde{\nu}}{\tilde{\nu}^4} = \int_0^\infty \epsilon(\lambda)f'(\lambda)\lambda^4 \, d\lambda \tag{2}$$

with $\varepsilon(\tilde{\nu}) = \varepsilon(\lambda)$ the molar absorbance of the acceptor at frequency $\tilde{\nu} = 1/\lambda$, $f(\tilde{\nu})[f'(\lambda)]$ the normalized fluorescence spectrum of the donor defined by

$$f(\tilde{\nu})\ d\tilde{\nu} = \frac{F(\tilde{\nu})\ d\tilde{\nu}}{\int_0^\infty F(\tilde{\nu})\ d\tilde{\nu}} = \frac{F'(\lambda)\ d\lambda}{\int_0^\infty F'(\lambda)\ d\lambda} = f'(\lambda)\ d\lambda \qquad (3)$$

where $F(\tilde{\nu})$ and $F'(\lambda)$ represent emission intensities per unit frequency and wavelength, respectively, n the refractive index of the medium in the range of overlap, and κ^2 the orientation factor for dipole-dipole coupling which is discussed below. R_o is a characteristic distance which is defined as the donor-acceptor separation for which the transfer rate equals the donor de-excitation rate in the absence of an acceptor:

$$R_o{}^6 = 8.785 \times 10^{-25}\kappa^2\phi_D n^{-4} J(\tilde{\nu})\ \text{cm}^6 \qquad (4)$$

The predicted dependence of the transfer rate on the inverse sixth power of R has been elegantly demonstrated experimentally [9, 11,12], as has that of the spectral overlap integral [13,14]. However, as has been pointed out [15,16], two experimental parameters, n and κ^2, have received relatively little attention, and usually some average value is assumed for them. While the error introduced by the uncertainty in the refractive index is not of great importance, that in the orientation factor may introduce large errors in R since κ^2 may theoretically take any value between 0 and 4.

2. Orientation Factor

The orientation factor is defined in terms of the angles between the unit vectors $\underset{\sim}{D}$, $\underset{\sim}{A}$, and $\underset{\sim}{R}$ which lie along the donor and acceptor transition moments and the direction joining them, respectively:

$$\begin{aligned}
\kappa &= \underset{\sim}{D} \cdot \underset{\sim}{A} - 3(\underset{\sim}{D} \cdot \underset{\sim}{R})(\underset{\sim}{A} \cdot \underset{\sim}{R}) \\
&= \cos \theta_T - 3 \cos \theta_D \cos \theta_A \\
&= \sin \theta_D \sin \theta_A \cos \varphi - 2 \cos \theta_D \cos \theta_A
\end{aligned} \qquad (5)$$

the angular relations being given in Fig. 1. Customarily, the ran-
dom solution average over all orientations of both $\underset{\sim}{D}$ and $\underset{\sim}{A}$ for which
$<\kappa^2> = 2/3$ is inserted wherever κ^2 appears. This is justified in
general only when it can be shown or reasonably assumed that the
relative orientations of D and A have a random distribution and that
the orientations are completely averaged within the lifetime of the
transfer process. It should be noted that the value $\kappa^2 = 0.476$
sometimes quoted for _rigid_ random orientations (highly viscous solvents or glasses as opposed to the mobile solvents implied above)
[4,17] is not generally appropriate, arising as it does (as $<\kappa> =$
0.690) in an approximation describing the concentration quenching
of donor emission by a statistically randomly distributed ensemble
of acceptors in nonmobile media [18-20].

3.   Transfer Efficiency

The efficiency of energy transfer for a D-A pair in a particular configuration i is given by the ratio of the transfer rate to
the rate of all deactivation processes:

$$T_i = \frac{k_{Ti}}{\tau_D^{-1} + k_{Ti}} \tag{6}$$

where k_{Ti} is given by Eq. (1). When D and A have some degree of
mutual orientational freedom, the transfer efficiency will be an
appropriate average value over all the possible configurations i.

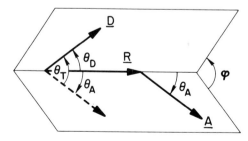

FIG. 1.   Geometry of a donor-acceptor transition moment vector
pair defining the dipole-dipole orientation factor κ^2 and the transfer depolarization factor d_T.

Two limiting averaging regimes for the transfer efficiency are of interest. When the relative motion of D and A is rapid compared with the donor decay rate so that all configurations of each D-A pair are sampled during the donor lifetime, the transfer efficiency for each D-A pair is determined by the dynamic average of the transfer rate over all allowed conformations $<k_{Ti}>$ as

$$<T>_d = \frac{<k_{Ti}>}{\tau_D^{-1} + <k_{Ti}>} \tag{7}$$

where

$$<k_{Ti}> = \sum_i n_i k_{Ti} / \sum_i n_i \tag{8}$$

n_i being the frequency with which a given configuration i occurs. It is useful to rewrite $<T>_d$ in terms of the D-A separation R and the characteristic distance R_o by substitution of Eq. (1) into Eq. (7) when

$$<T>_d = \frac{<R_o^6>/R^6}{1 + (<R_o^6>/R^6)} \tag{9}$$

where $<R_o^6>$ is obtained by inserting $<\kappa^2>$, the average value of the orientation factor over all configurations, into Eq. (4). Since in the dynamic limit the same $<T>_d$ characterizes each D-A pair, it is also the average value of the transfer efficiency for the whole ensemble of D-A pairs.

On the other hand, in static averaging, while all configurations i are represented in the ensemble of D-A pairs, that of any particular pair does not change during the donor lifetime. Each pair contributes a particular transfer efficiency T_i to the overall efficiency which is thus a static average $<T>_s$ over the ensemble:

$$<T>_s = \sum_i n_i T_i / \sum_i n_i \tag{10}$$

or, by using Eq. (6)

$$<T>_s = < \frac{k_T}{\tau_D^{-1} + k_T} > \tag{11}$$

Under dynamic averaging conditions all allowed relative orientations of each donor-acceptor pair are sampled. Since transfer is more likely to occur when the relative orientations are favorable and no more orientations are sampled once transfer has occurred, the dynamically averaged transfer efficiency $<T>_d$ always exceeds the statically averaged one $<T>_s$. This statement $(<T>_d \geq <T>_s)$ is, according to Eqs. (7) and (11), equivalent to

$$\frac{<x>}{1 + <x>} \geq <\frac{x}{1 + x}>$$

where $x_i = k_{Ti}\tau_D$, and each average is taken over all possible configurations i. Since the sum over N terms can be taken two terms (i and j) at a time, the statement is true if

$$\frac{x_i + x_j}{2 + x_i + x_j} \geq \frac{1}{2} \left(\frac{x_i}{1 + x_i} + \frac{x_j}{1 + x_j} \right)$$

i.e., if

$$x_i^2 + x_j^2 \geq 2x_i x_j$$

which holds for all positive x_i and x_j since $(x_i - x_j)^2 \geq 0$. The magnitude of the difference between $<T>_d$ and $<T>_s$ depends on the motional freedom enjoyed by A and D as well as on $<T>$ itself. It vanishes as the transfer efficiency approaches zero, or when A and D are uniquely oriented with respect to each other, in which case each pair remains in this particular fixed conformation during the donor lifetime.

4. Experimental Considerations

On macromolecular substrates such as proteins or nucleic acids, one or both of $\underset{\sim}{D}$ and $\underset{\sim}{A}$ may be rigidly fixed with respect to $\underset{\sim}{R}$, and, if some orientational freedom is allowed, it is generally restricted.

Under normal aqueous solution conditions, dynamic averaging over any
such freedom can be expected, although its randomness over the restricted range may be questionable.

Obviously, it would be desirable to have some method for determining the appropriate value of κ^2 or its average, since, although
higher actual values (up to 4) result in relatively little change of
the R value eventually determined by using an R_o calculated with the
assumption that $<\kappa^2> = 2/3$, the lower values (near zero) can lead
to large errors (see Fig. 2). Use of several D-A pairs for which
different mutual orientations are ensured by varying the way in which
they are bound to their specific sites on the macromolecule has been
invoked to help eliminate uncertainty in the orientation factor [3,
15], and it has also been suggested that if transfer takes place into
overlapping acceptor absorption bands of different polarization, κ^2
is again experimentally derivable [16]. As will be seen later, the
latter alone can only reduce the uncertainty in $<\kappa^2>$ to the range
of values 0 to 2. In a special case where the acceptor was a cobalt
ion whose absorption in the overlap region was made up of slightly
split components of a triply degenerate transition corresponding to
an isotropic three-dimensional oscillator [5], the uncertainty in
the orientation factor was reduced to $1/3 \leq \kappa^2 \leq 4/3$, giving an uncertainty in derived separation values R of

$$(0.5)^{1/6}R_{2/3} \leq R \leq (2)^{1/6}R_{2/3}$$

i.e., approximately ±12%, where $R_{2/3}$ is the value obtained using
$<\kappa^2> = 2/3$. While x-ray structural determinations may in principle
provide information on κ^2, e.g. [21], this approach is often impractical and deals moreover with the crystalline system for which an
intramolecular distance determination has been obviated once the
x-ray structure has been solved, except for a comparison of the solution and crystalline configurations. On the other hand, the angle
between the donor absorption and acceptor emission vectors can be
determined from the depolarization of fluorescence excited with light
absorbed by the donor and emitted by the acceptor. The feasibility

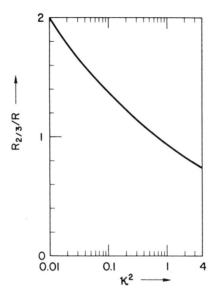

FIG. 2.   Ratio of apparent to true donor-acceptor separation obtained from an energy transfer efficiency as a function of the actual
orientation factor K^2, the apparent separation having been computed
on the assumption that $K^2 = 2/3$, the dynamic random average value.

of using the technique of fluorescence depolarization to estimate
the possible range of values of K^2 or its average is explored in the
following sections.

B.   Fluorescence Depolarization

1.   Emission Anisotropy

The absorption and subsequent emission of a photon are highly
polarized processes. The probability of electromagnetic radiation
with a given electric vector polarization interacting with a suitable
absorber (which may be represented most simply by the dipole transition moment vector associated with a virtual linear oscillator) and
the subsequent release of a quantum of electromagnetic radiation in
a particular direction is a function not only of the relative orien-

tations of the electric vectors and propagation directions of the
excitation and emission light but also of the transition moment vector with respect to these [22]. As a result, the fluorescence of
solutions of an absorber-emitter is expected to be polarized to some
extent when viewed from an arbitrary direction.

The most useful measure of this polarization is the emission
anisotropy (EA), r, introduced by Jabłoński [23,24]. It may be defined as

$$r = \frac{I_V - I_H}{I} \tag{12}$$

where I_V, I_H are the vertically and horizontally polarized intensities of emission observed in a horizontal plane and at right angles
to the direction of arbitrarily polarized excitation incident on the
sample. I corresponds to the total emission intensity, i.e., the
sum of the emission intensities polarized along any three mutually
orthogonal axes. The excitation and observation conditions of the
greatest use experimentally are given in Appendix A along with some
of the fundamental equations of polarized emission spectroscopy and
the relationships between EA and the still commonly used classical
degree of polarization, p. The remainder of this chapter will be
formulated in terms of emission anisotropy, since this quantity is
conceptually more useful than the degree of polarization and the
majority of relationships involved are more simply derived and expressed.

2. Useful Relationships

A number of the relationships derived in Appendix A and most
pertinent to the material of this chapter are summarized here:

(i) The EA is measured in laboratory coordinates according to
Eqs. (A19a)-(A19c) as

$$r = \frac{I_V - I_H}{I_V + 2I_H} \quad \text{(vertically polarized excitation)} \tag{13a}$$

or

$$r_n = \frac{I_V - I_H}{2I_V + I_H} = \frac{r}{2} \quad (\text{"natural" excitation}) \tag{13b}$$

(ii) The average EA observed after a series of independent isotropic depolarizing events may be expressed [compare Eqs. (A6), (A7), and (A22)] as a product of azimuthally averaged depolarization factors $<d_i>$ and the EA of the isolated luminophore in the absence of motion which is called the <u>fundamental</u> EA or r_f:

$$<r> = r_f \prod_i <d_i> = r_f \prod_i (\frac{3}{2} <\cos^2 \theta_i> - \frac{1}{2}) \tag{14}$$

Here each depolarization stage i contributes a factor described by the average cosine square of the orientation change θ_i. In particular, isotropic Brownian rotational movements or energy transfer may be responsible for such depolarizing factors.

(iii) The average EA of a number of classes of emitter distinguished by different EA values is given by the simple addition law of Eq. (A25a):

$$<r> = \sum_{i=1}^{N} r_i I_i / \sum_{i=1}^{N} I_i \tag{15}$$

where I_i represents the <u>total</u> emission intensity of the ith species. Inversely, this law may be used to analyze the excitation spectrum (and by implication the absorption spectrum) of such a system, be it due to a mixture of separate luminophores or the overlapping of different transitions of a single luminophore.

C.   Transfer Depolarization and the Orientation Factor

1. Definition of Transfer Depolarization

Suppose that a donor (D) and acceptor (A) attached to a macro-molecular substrate can undergo one and only one transition each in

the energy region under consideration, that the donors have their
absorption transition moment vector directions distributed isotropically in the sample solution, and that they are excited by vertically polarized light.

Immediately following absorption of photons, the distribution
of excited transition moments has a maximum in the vertical direction and the emission anisotropy r_f is 0.4 (see Appendix A). If the
excited donors (D*) are more or less free to move with respect to
the substrate, their EA is reduced, according to the isotropic depolarization law (Appendix A), by a factor

$$<d_D'> = (\frac{3}{2} < \cos^2 \theta_D' > - \frac{1}{2})$$

where the average is taken over the motion of all excited donors,
and θ_D' is the angle by which the transition moment of D changes its
direction relative to the substrate during this motion. The appli-
cability of the isotropic depolarization law depends on θ_D' having
an azimuthally uniform distribution about the direction of the absorption moment vector at the instant of excitation.

If, on the other hand, energy transfer occurs, the excited acceptors have an EA analogously reduced by a transfer depolarization
factor to

$$r_{A(D)} = r_{fD}(\frac{3}{2} < \cos^2 \theta_T > - \frac{1}{2}) = r_{fD}<d_T> \tag{16}$$

where θ_T is the angle between the D and A transition moments, and
the average is taken over all mutual orientations of donor and acceptor covered on average during the growth and decay of the excited
acceptor population. Such a value is to be compared with the isotropic depolarizations of directly excited D and A which may be
written

$$r_D = r_{fD}(\frac{3}{2} < \cos^2 \theta_D' > - \frac{1}{2}) = r_{fD}<d_D'> \tag{17a}$$

and

$$r_A = r_{fA}(\frac{3}{2} < \cos^2 \theta_A' > - \frac{1}{2}) = r_{fA} < d_A' > \qquad (17b)$$

where θ_D', θ_A' are the orientation changes of D and A moment vector directions relative to the substrate after excitation, and their cosine square averages are over the decays of excited donor and acceptor populations, respectively. As indicated in the previous section, analogous equations hold for <u>natural</u> excitation.

This simple picture becomes more complex in general when rotation of the macromolecule itself is also considered. However, it is usually possible to separate out the effect of macromolecular motion experimentally (see Sec. IV and Appendix C).

2. Experimental Considerations

If the donor and acceptor absorption and fluorescence bands do not overlap, $r_{A(D)}$ and r_A, as well as r_D, are readily determined from the relevant polarization excitation spectra. In the presence of overlapping D and A absorption spectra the contribution of each band in the fluorescence excitation spectrum to the observed EA is simply determined by application of Weber's addition law (see Sec. II,B,2 and Appendix A):

$$r_{obs} = \frac{r_A I_A + r_{A(D)} I_{A(D)}}{I_A + I_{A(D)}} \qquad (18)$$

for a given excitation wavelength, where I_A is the emission intensity of the acceptor excited directly and $I_{A(D)}$ that excited in the donor band. Since I_D and $I_{A(D)}$ have the same dependence upon excitation wavelength in this case, and it is known how I_A and I_D vary with wavelength from their absorption spectra, and since r_A can be determined either in a region where D does not absorb or in a separate experiment, $r_{A(D)}$ is fully determined. If D also contributes some fluorescence at the wavelength of observation, this may be similarly allowed for; for a given excitation wavelength:

$$r_{obs} = \frac{r_D I_D + r_A I_A + r_{A(D)} I_{A(D)}}{I_D + I_A + I_{A(D)}} \qquad (19)$$

The situation in the presence of more than one donor and acceptor level is a little more complex. In principle, the same kind of analysis is carried out, but r_D and r_A are now wavelength dependent rather than essentially constant, since the transition moment vectors of the upper levels are usually oriented at some finite angle with respect to those of the lowest levels. Likewise, $r_{A(D)}$ cannot be expected to be constant. On the other hand, having been extracted from the experimental data as a transfer polarization excitation spectrum, it will still be related to the donor absorption (or fluorescence excitation) spectrum by the addition law, e.g., for two D levels, 1 and 2:

$$r_{A(D)} = \frac{r_{A(D_1)}I_1 + r_{A(D_2)}I_2}{I_1 + I_2} \tag{20}$$

at a given excitation wavelength, where $r_{A(D_1)}$ and $r_{A(D_2)}$ are the EA's for transfer from the first and second donor levels (by transfer from D_2 is meant internal conversion to D_1 followed by transfer from the latter) and are essentially independent of excitation wavelength, while I_1 and I_2 are proportional to the polarized absorption spectra of these two transitions.

If the transitions are not well-separated so that regions of constant $r_{A(D_1)}$ and $r_{A(D_2)}$ are not evident, the transfer EA's must be determined by fitting the derived transfer polarization spectrum and donor absorption spectrum to limiting values of $r_{A(D_1)}$ and $r_{A(D_2)}$ [one greater than the highest $r_{A(D)}$ observed, the other smaller than the lowest; $r_{A(D_2)}$ is not necessarily less than $r_{A(D_1)}$]. The derived polarized D_1, D_2 absorption spectra should coincide with those obtained by a similar analysis of the donor excitation spectra:

$$r_D = \frac{r_{D_1}I_1 + r_{D_2}I_2}{I_1 + I_2} \tag{21}$$

where r_{D_1} and r_{D_2} are the EA's of donor emission (here $r_{D_2} \le r_{D_1}$ always obtains).

The kinds of absorption, excitation, and polarization excitation spectra to be expected for donor-acceptor pairs with sufficient

spectral overlap to lead to measurable energy transfer are shown
schematically in Fig. 3. In the situation depicted, transfer is
possible only to the lowest acceptor level, and only the acceptor
emission is monitored.

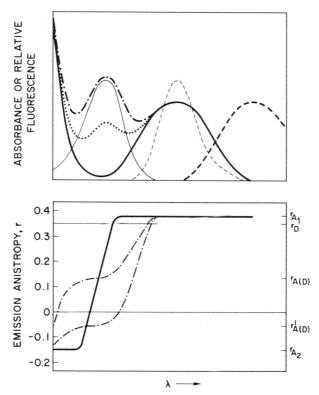

FIG. 3. Upper: Absorption, excitation, and emission spectra of
a donor-acceptor pair. The light and heavy curves refer to D and A,
respectively: solid line, absorption spectra of isolated D and A;
dashed line, emission spectra of D and A; dot-dashed line, absorption
spectrum of the D-A pair; dotted line, excitation spectrum for A
emission in the presence of transfer; when the transfer efficiency
is unity, this spectrum coincides with the absorption spectrum of
the D-A pair. Lower: Polarization excitation spectra of isolated
A (heavy solid line), isolated D (solid line), and of two representative D-A pairs (dot-dashed line) with different relative orientations: r_{A_1} and r_{A_2} are the EA's of the first and second acceptor
absorption bands, r_D the donor EA and $r_{A(D)}$, $r'_{A(D)}$ are the two transfer EA's excited exclusively via the donor.

3. Relationship between κ^2 and d_T

The analyses outlined above apply whether or not transfer takes place partially or entirely into the lowest acceptor transition, since the observed depolarization depends only on the mutual orientations of the initial absorption and final emission vector directions. However, of ultimate concern to this discussion is the orientation factor in transfer which is determined by the angular relationships between the D emission moment vector and the A absorption moment vector. Since the extent of transfer depolarization is related to the D absorption and A emission vectors, it can be meaningfully related to the orientation factor only when transfer takes place essentially entirely into the lowest A absorption band and excitation is in the lowest D absorption band. Consequently, cases of transfer into more than one acceptor band will be excluded from the rest of the discussion and only the $r_{A(D)}$ for the lowest donor band considered further, it being assumed that any necessary analysis may be carried out in the manner indicated to yield the appropriate transfer EA from which the transfer depolarization factor d_T or its average is obtainable.

From Eq. (5) it is immediately obvious that, in general, it is not possible to determine a specific or average value of κ^2 directly from a transfer depolarization factor but that the depolarization measurement sets some limitations on the possible values of κ^2 or its average. This is more clearly seen if $\cos \theta_T$ is rewritten for comparison with κ^2, Eq. (5), as

$$\cos \theta_T = \sin \theta_D \sin \theta_A \cos \varphi + \cos \theta_D \cos \theta_A \qquad (22)$$

Thus, if, for example, D and A are both rigidly oriented with respect to each other on the macromolecular substrate, κ^2 is fully determined by $\cos^2 \theta_T$ if $\varphi = \pi/2$ when, according to Eqs. (5) and (22), it is equal to $4 \cos^2 \theta_T$ and least determined for $\varphi = 0$ and $\varphi = \pi$ ($0 \le \theta_{A,B} \le \pi/2$). The graphical comparisons of Fig. 4 demonstrate that, with no knowledge of φ

$$\kappa^2_{min} = 0 \qquad\qquad\qquad\qquad\qquad (23a)$$

for all d_T and

$$\frac{9}{4} \leq \kappa^2_{max} = \frac{1}{4}\left[3 + \cos\theta_T\right]^2 = \frac{1}{4}\left[3 + \left\{\frac{1}{3}\left(2d_T + 1\right)\right\}^{1/2}\right]^2 \leq 4 \qquad (23b)$$

This case of rigid relative orientations of D and A leads to the maximum uncertainty in κ^2, setting merely a high upper limit to a derived D-A separation, R. Usually, however, D and/or A have some degree of freedom with respect to the substrate, and then appropriate average values of κ^2 and d_T should be employed. While, in general, it is still impossible to determine the unique average value of the orientation factor $<\kappa^2>$ from the average transfer depolarization factor $<d_T>$, it turns out that the range of possible values of $<\kappa^2>$ is often considerably reduced when its limits have to be consistent with observed $<d_T>$ and $<d'>$ values. The reason for this will be demonstrated in the following section by the analysis of a variety of specific models of restricted freedom of D and/or A. For the purposes of the argument, only analytically solvable limiting cases involving either one or two angular variables will be examined here. More complete results for some of the models including, in particular, the effect of variable transfer efficiency in the static averaging regime will be presented elsewhere.

In the above, only the case of transfer between a single donor and a different acceptor has been considered, the acceptor furthermore being incapable of transferring energy to the donor. The case of self-transfer thus excluded is less tractable than the above, since the depolarization of transferred excitation energy is the only means of actually observing that transfer is taking place, and the system is therefore underdetermined experimentally. However, the case of multiple self-transfers among a number of identical luminophores can provide some information in particular instances as described in [25-28] and also elsewhere in this volume [29].

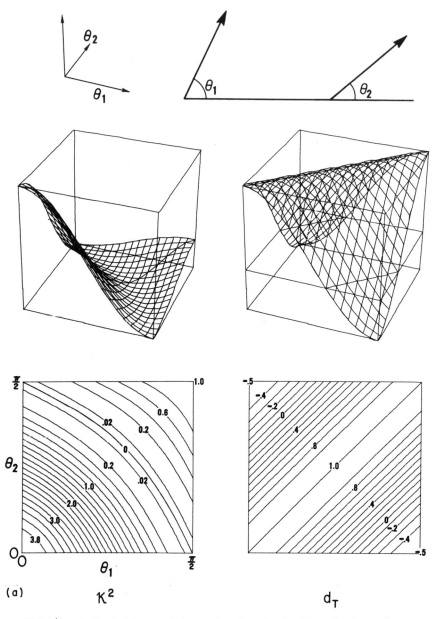

(a) κ^2 d_T

FIG. 4. Orientation and transfer depolarization factors for a
D-A pair with fixed relative orientations as functions of θ_1 and θ_2
with (a) $\varphi = 0$ and (b) $\varphi = \pi$ — see Eqs. (5), (16), and (22). The
labeled triads indicate the aspects of the three-dimensional plots.

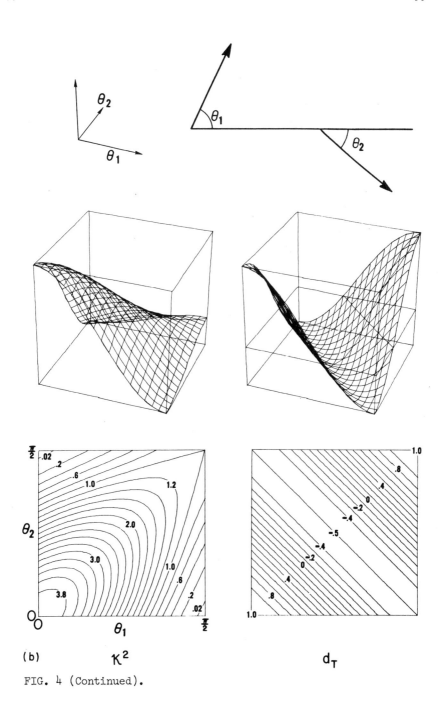

(b) κ^2 d_T

FIG. 4 (Continued).

Geometrical information has also been obtained by analyzing models of proteins labeled with several donors and a single acceptor [30] or several acceptors [31,32] and comparing these model results with experimental values for the average transfer efficiency and average donor emission lifetime only.

III. SELECTED MODELS

When a donor and acceptor attached to a substrate have some degree of freedom of mutual orientation, both dynamic and static conditions of averaging over this freedom are realizable in principle, independent of the motion of the substrate itself. In the latter regime, the transfer depolarization factor observed for the ensemble of stationary D-A pairs and defined as

$<d_T>_s = 3/2 <\cos^2 \theta_T>_s - 1/2$ [see Eqs. (14) and (16)] will be given by the average over the transfer efficiency for configurations i by

$$< \cos^2 \theta_T >_s = \sum_i n_i T_i \cos^2 \theta_{Ti} / \sum_i n_i T_i \qquad (24)$$

where n_i is the relative frequency and T_i the transfer efficiency for the configuration i and θ_{Ti} the corresponding angle between donor and acceptor moments. Under dynamic averaging conditions, however, the observed depolarization factor is given simply as its average value over the configurations

$$< \cos^2 \theta_T >_d = \sum_i n_i \cos^2 \theta_{Ti} / \sum_i n_i \qquad (25)$$

In the various models analyzed and presented here, n_i has been taken as a purely geometrical weighting element. The transition moments of the interacting luminophores are each supposed to take up random orientations either over the surface or volume of a cone, corresponding, e.g., to rotational freedom of the luminophore about a single bond in the first case or approximating to the result of rotational freedom about several bonds in the second (see further

discussion in Appendix C). This allows the replacement of the summations of Eqs. (24) and (25) by integration over the relevant geometrical elements. The dynamic average is thus obtainable analytically, while, in general, the static average is not. However, in
the limit in which $<T>$ approaches zero, the transfer rate becomes
negligible compared with that of donor emission, see Eq. (11), and
Eq. (24) is reduced to an analytically integrable form:

$$\lim_{<T>\to 0} <\cos^2 \theta_T>_s = \sum_i n_i \kappa_i^2 \cos^2 \theta_{Ti} / \sum_i n_i \kappa_i^2 \qquad (26)$$

In this limit the various configurations i are seen to be weighted
by κ_i^2 itself, whereas for finite values of T_i this weighting factor
is reduced, according to Eqs. (1) and (6), to $\kappa_i^2/[1 + (R_{0i}/R)^6]$ so
that $<d_T>_s$ always lies between $<d_T>_d$ and $\lim_{<T>\to 0} <d_T>_s$.

In order to illustrate the relationship between the transfer
depolarization factors and $<\kappa^2>$ which is expressed by

$$<\kappa^2> = \sum_i n_i \kappa_i^2 / \sum_i n_i \qquad (27)$$

integrations corresponding to the summations in Eqs. (25)-(27) were
performed for a variety of models of the kind indicated above.
These include various relative orientations of $\underset{\sim}{D}$ and $\underset{\sim}{A}$ with respect
to $\underset{\sim}{R}$, the unit vector joining them and allowing for distribution of
either or both of them over the surface or volume of a cone. A more
detailed description of the models is given in Appendix B together
with the integrated expressions for $<\kappa^2>$, $<d_T>_d$ and $\lim_{<T>\to 0} <d_T>_s$
as functions of either one or a pair of angular variables descriptive
of each particular model. Graphical representations of these expressions are given in Figs. 5-51. For those models in which the
average orientation factor and average transfer depolarizations are
functions of two angular variables, the results are displayed in the
form of three-dimensional graphs in true perspective and as contour
plots. In each model the two angular parameters describing the
fixed orientation or orientational freedom of the transition moment

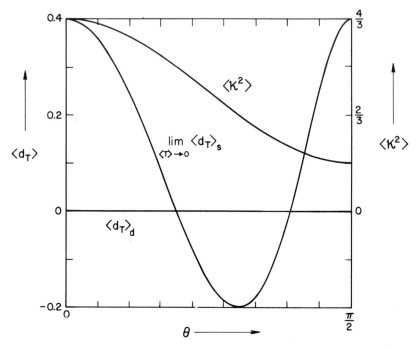

FIG. 5. Orientation and transfer depolarization factors for
Model 1 (see Appendix B for a detailed description of this and succeeding models).

vectors may be arbitrarily assigned to donor or acceptor whose roles
are completely interchangeable with respect to the three quantities
derived from the models.

The first two examples represent models defined by <u>single</u> angular variables. In Fig. 5, for instance, the model is that of one
transition moment fixed, the other free to take up any orientation
within a sphere or, perhaps more realistically, within a hemisphere
whose symmetry axis coincides with $\underset{\sim}{R}$. The model described in Fig. 6
has $\underset{\sim}{D}$ and $\underset{\sim}{A}$ constrained to take up any orientation in planes which
are at a fixed angle φ to each other.

The remaining models defined by pairs of angular variables com-
prise examples in which one of the transition moment vectors is
rigidly oriented with respect to $\underset{\sim}{R}$, the other being constrained to

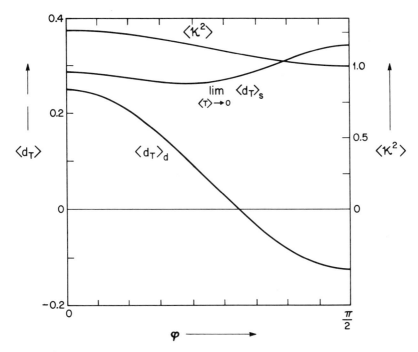

FIG. 6.   Orientation and transfer depolarization factors for Model 2.

lie anywhere on the surface or within the volume of a cone (Figs. 7-16), or both vectors are so constrained. Included in the latter are cases where one cone axis is either parallel to or in quadrature with $\underset{\sim}{R}$ and the other axis (Figs. 8-10, 17-40), as well as cases in which the vectors are confined to planes which may be thought of as corresponding physically to the intercalation planes of some polymers, e.g., nucleic acids, polyphosphates (Figs. 41-45). Finally, a set of cases is presented for which one transition moment vector is rigidly oriented with respect to $\underset{\sim}{R}$, the other being constrained to lie on the surface or within the volume of a cone, the orientation of whose axis is determined by that of the first vector in such a way as to give a minimum or maximum value of $<\kappa^2>$ for a given cone half-angle (Figs. 46-51) — compare the rigid-rigid cases in the previous section (Fig. 4).

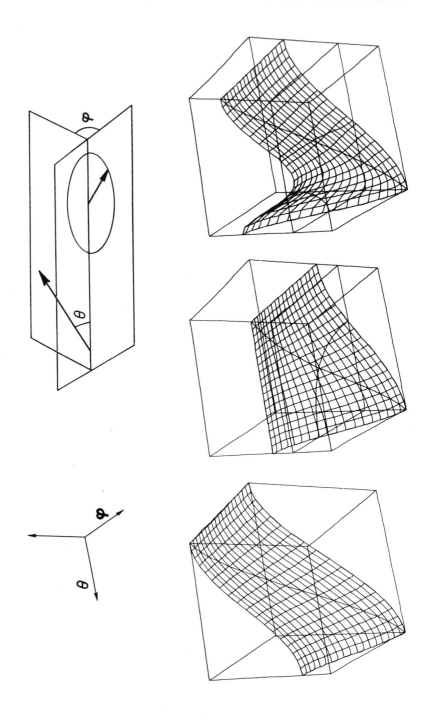

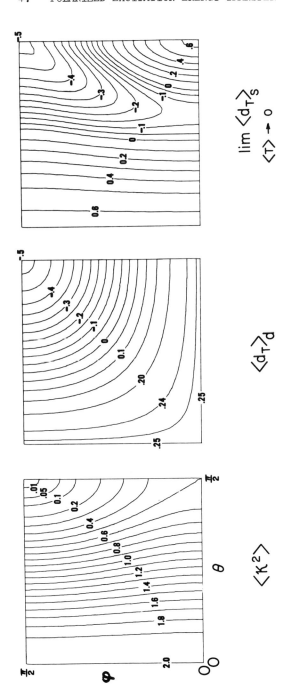

FIG. 7. Orientation and transfer depolarization factors for Model 3. The labeled triad defines the aspect of the three-dimensional plots as does the corresponding triad in each of the succeeding models.

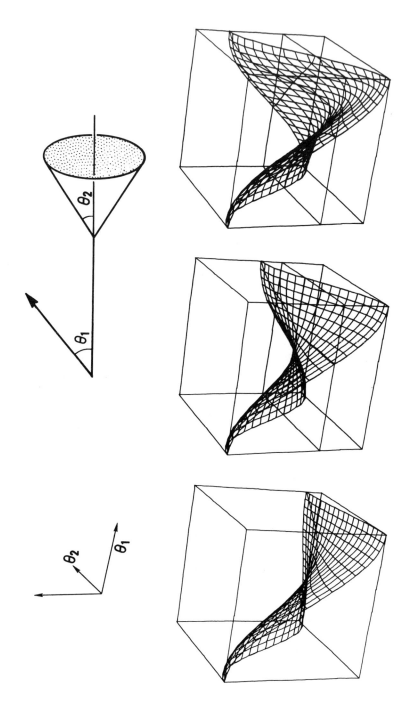

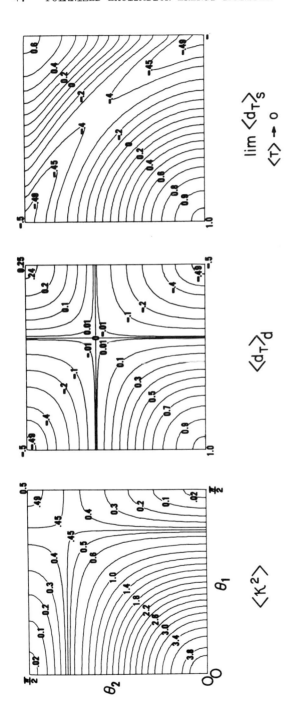

FIG. 8. Orientation and transfer depolarization factors for Model 4(i,c) and 5(i,cc).

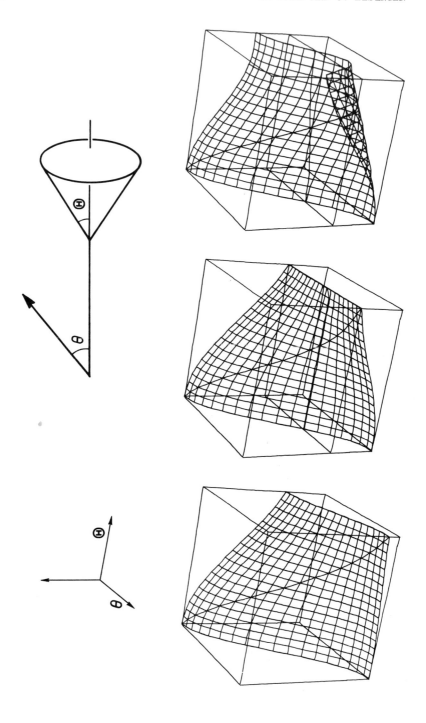

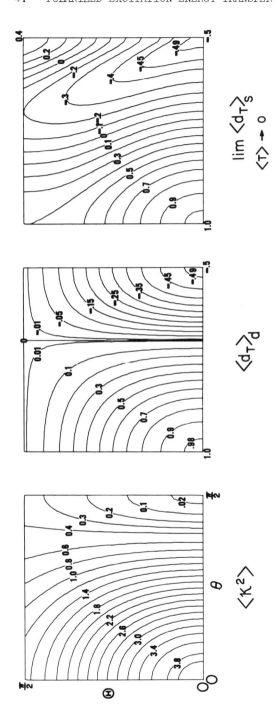

FIG. 9. Orientation and transfer depolarization factors for Model 4(i,C) and 5(i,cC).

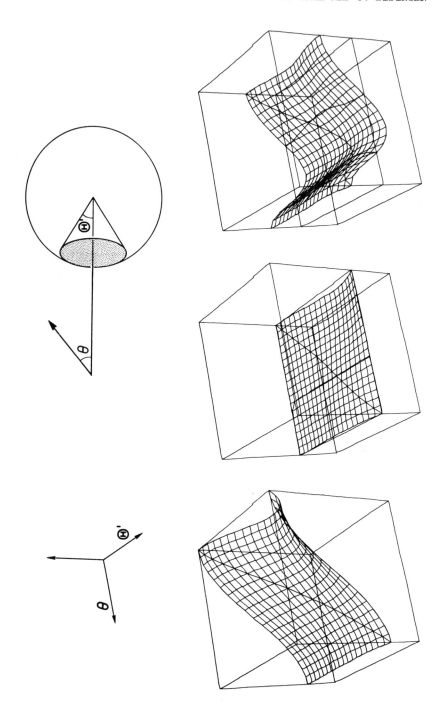

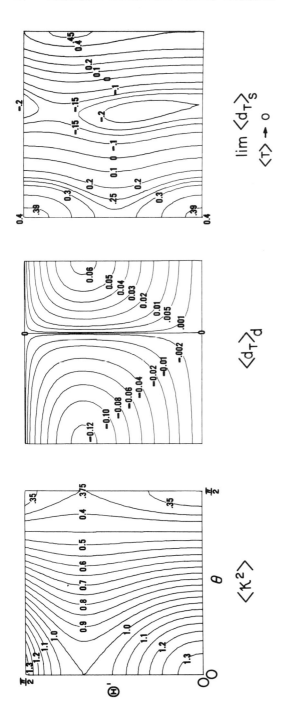

FIG. 10. Orientation and transfer depolarization factors for Model 4(i,c') and 5(i,cC').

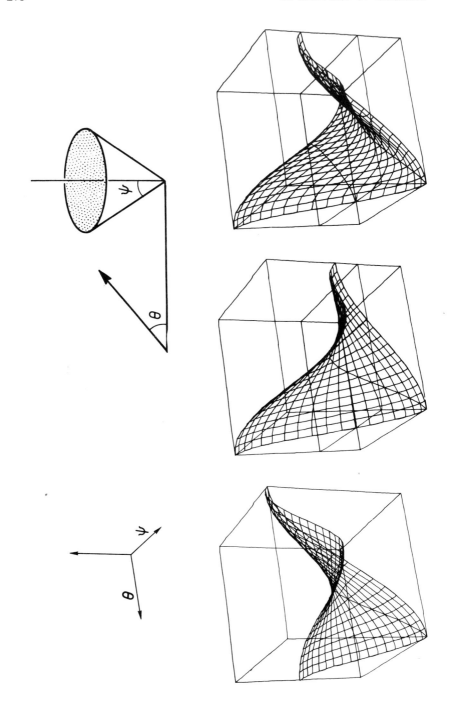

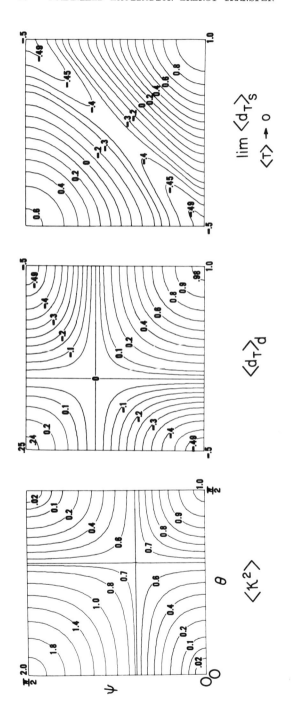

FIG. 11. Orientation and transfer depolarization factors for Model 4(ii,c).

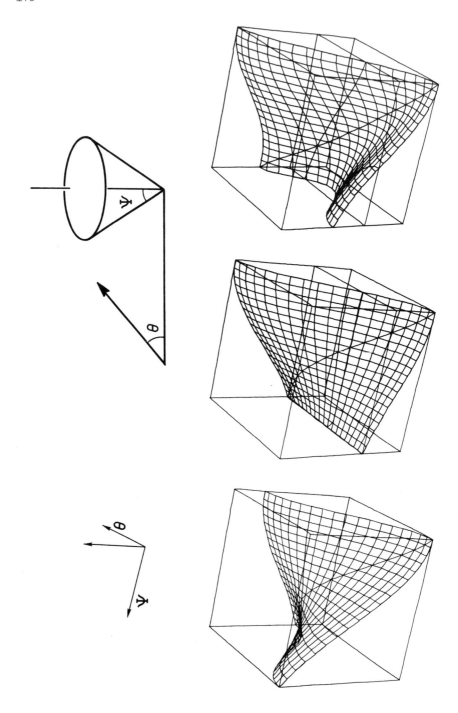

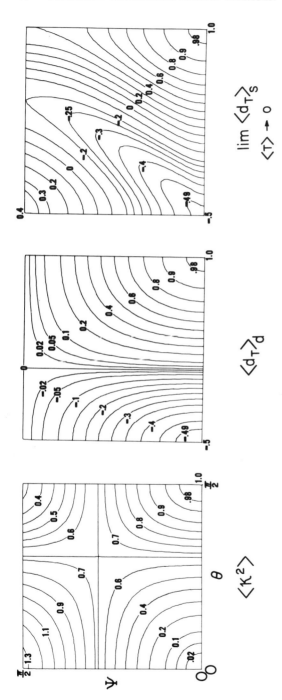

FIG. 12.   Orientation and transfer depolarization factors for Model 4(ii,C).

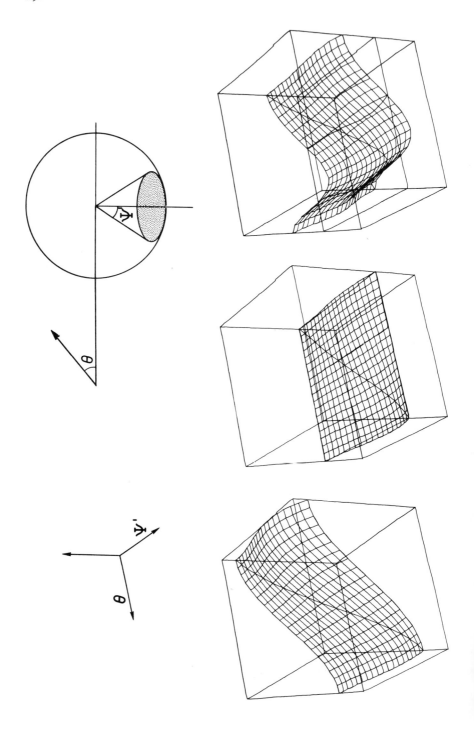

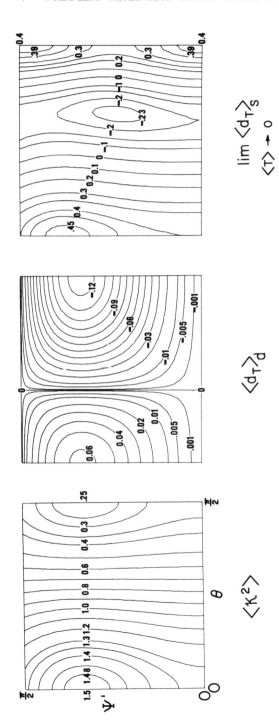

FIG. 13. Orientation and transfer depolarization factors for Model 4(ii,C').

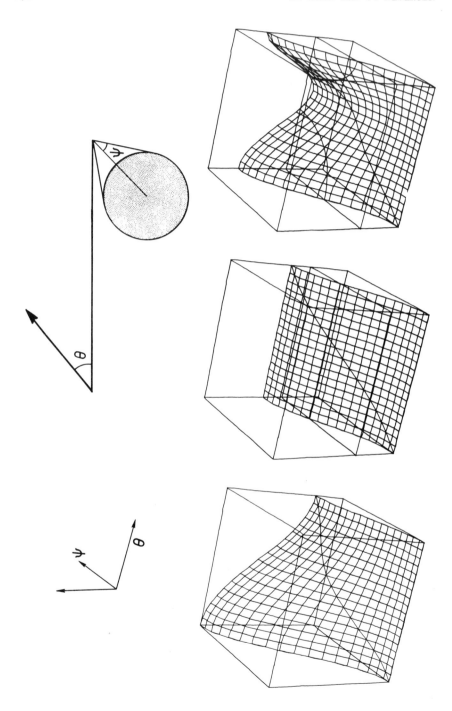

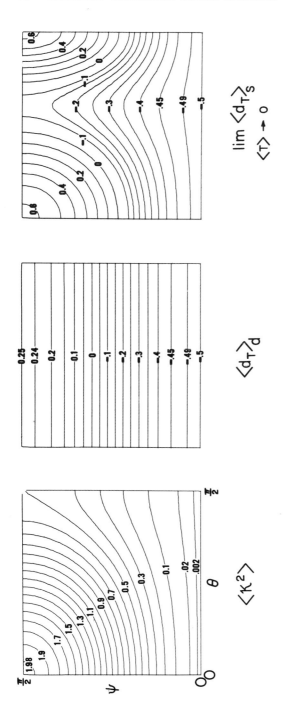

FIG. 14. Orientation and transfer depolarization factors for Model 4(iii,c).

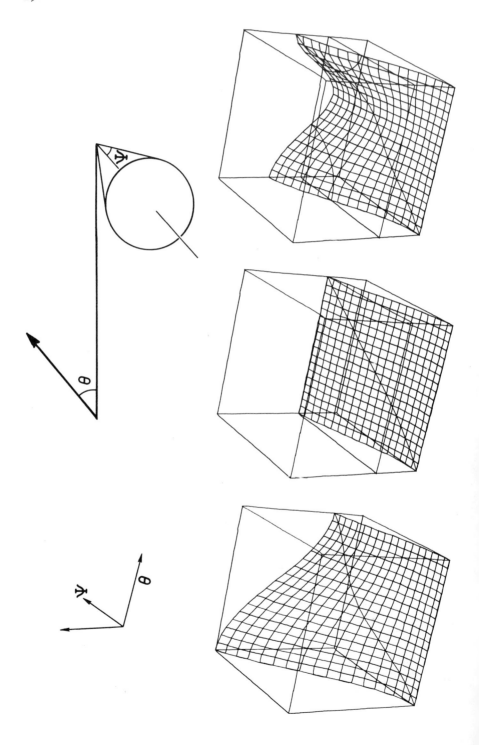

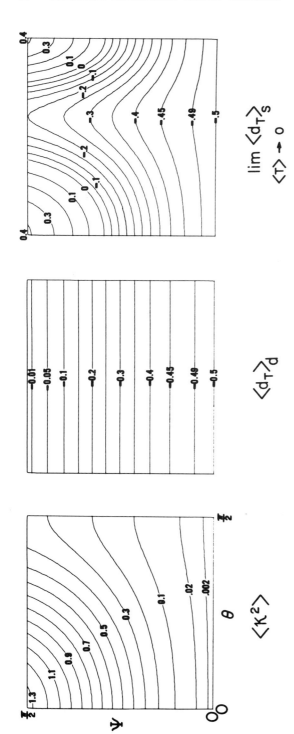

FIG. 15. Orientation and transfer depolarization factors for Model 4(iii,C).

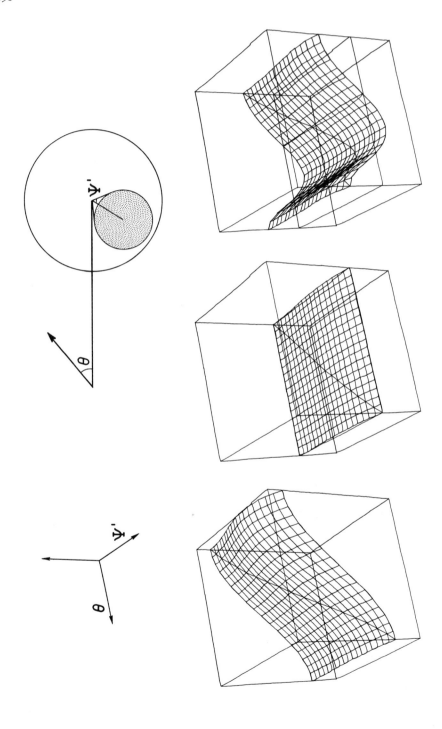

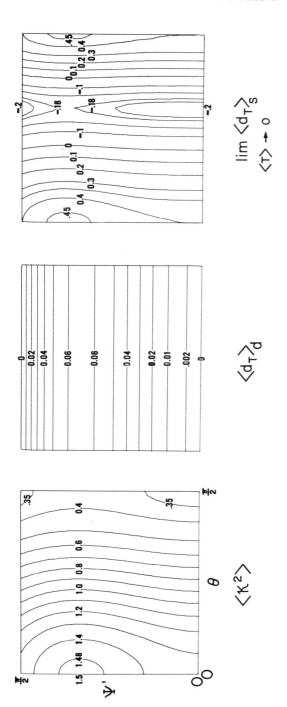

FIG. 16. Orientation and transfer depolarization factors for Model 4(iii,C').

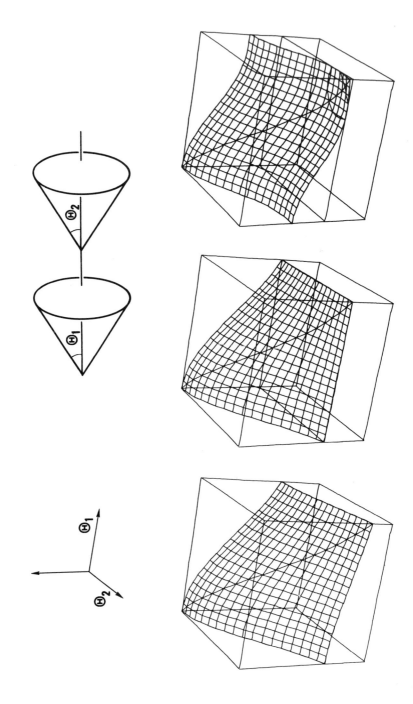

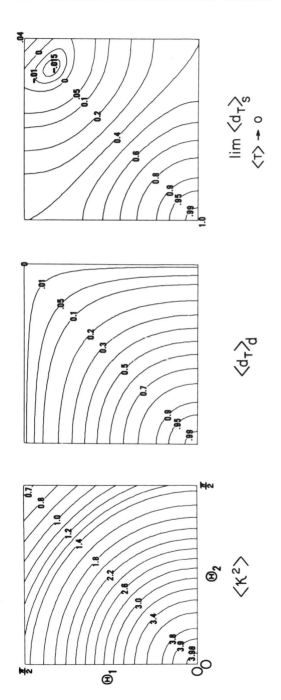

FIG. 17. Orientation and transfer depolarization factors for Model 5(i,CC).

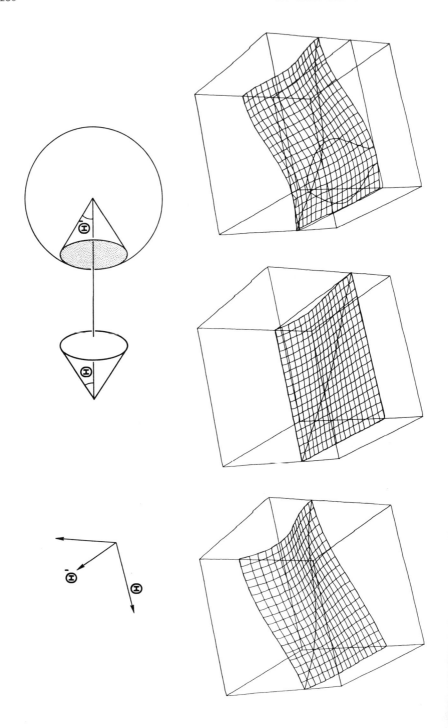

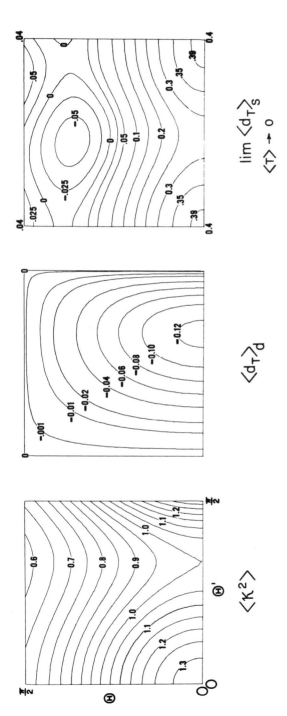

FIG. 18.   Orientation and transfer depolarization factors for Model 5(i,CC').

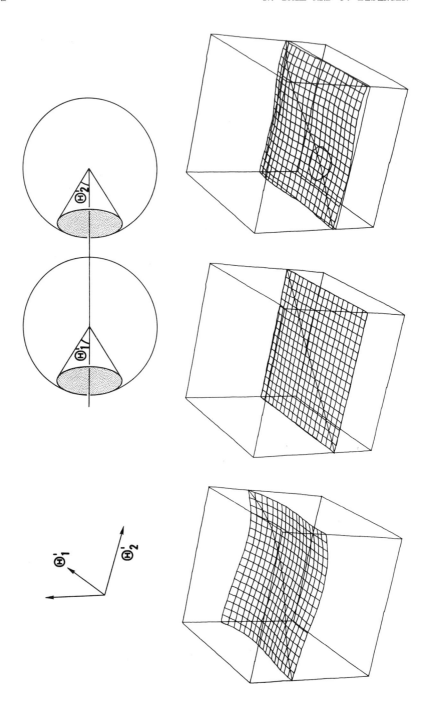

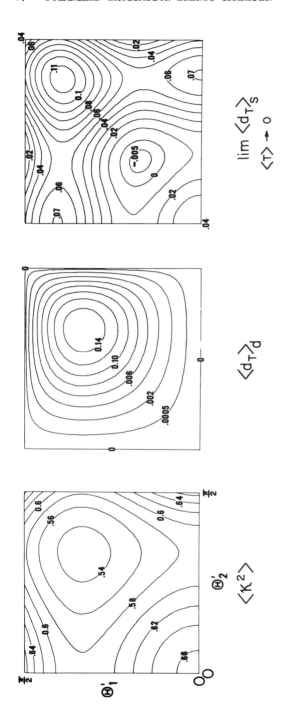

FIG. 19. Orientation and transfer depolarization factors for Model 5(i,C'C').

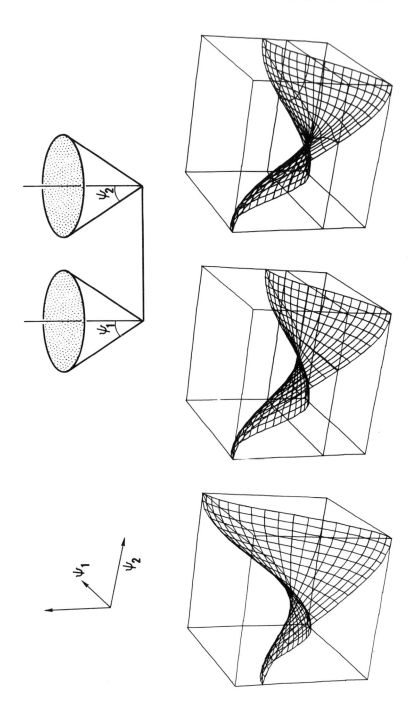

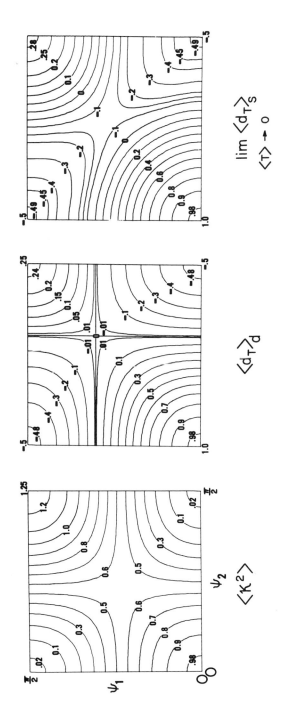

FIG. 20.   Orientation and transfer depolarization factors for Model 5(ii,cc).

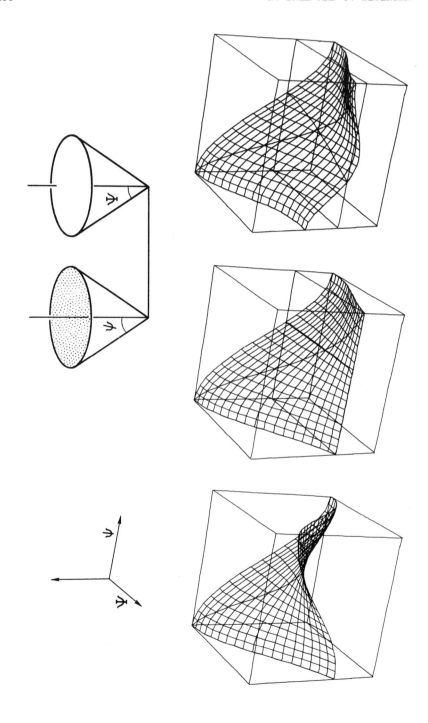

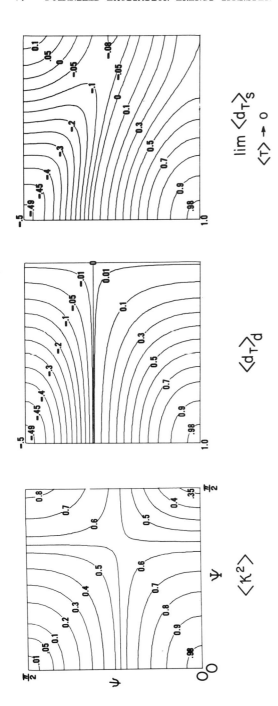

FIG. 21.   Orientation and transfer depolarization factors for Model 5(ii,cC).

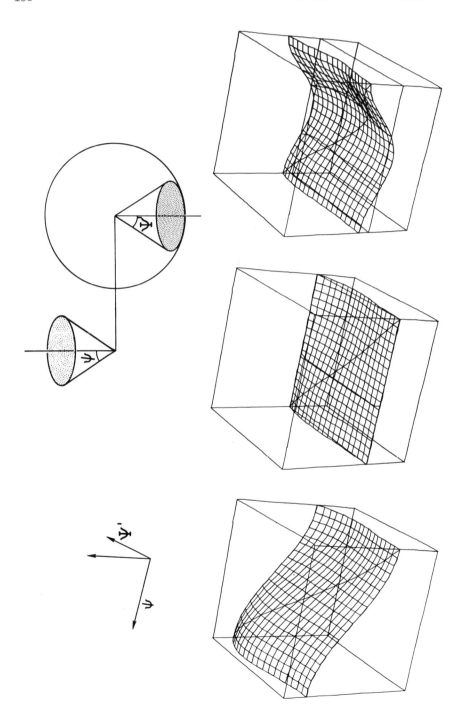

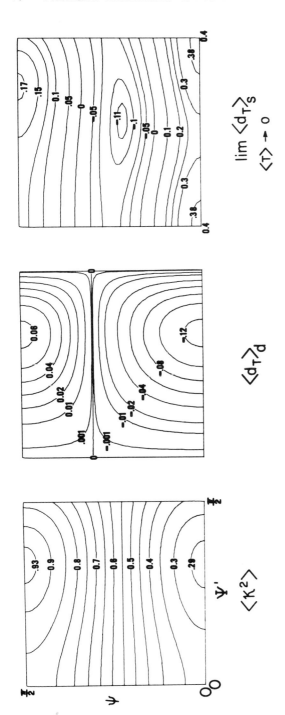

FIG. 22.   Orientation and transfer depolarization factors for Model 5(ii,cC′).

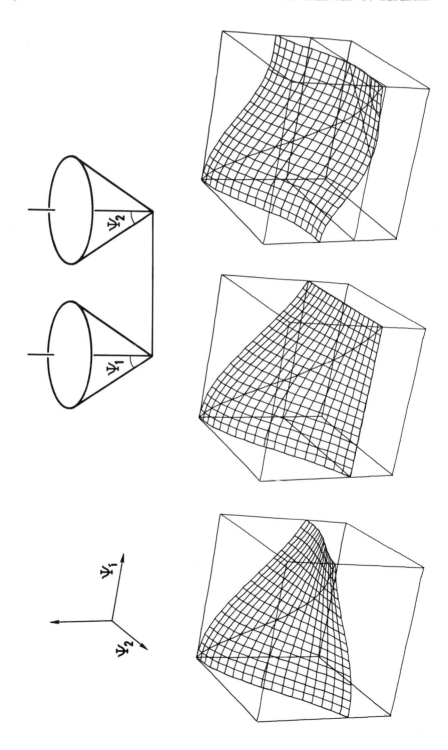

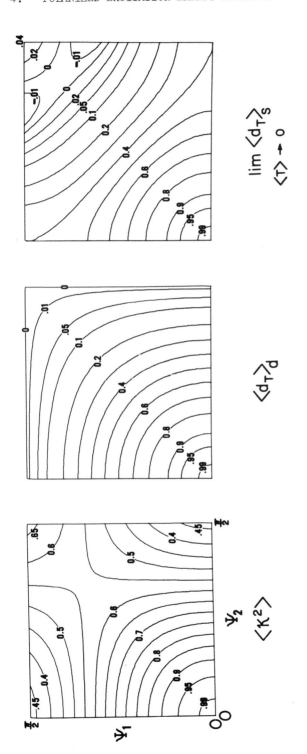

FIG. 23).   Orientation and transfer depolarization factors for Model 5(ii,CC).

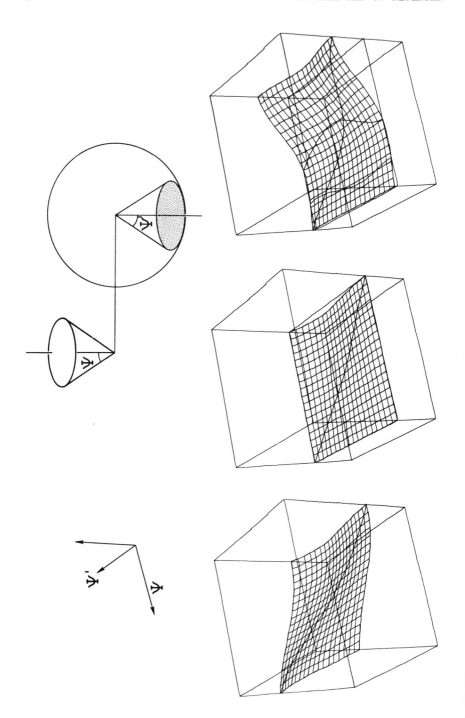

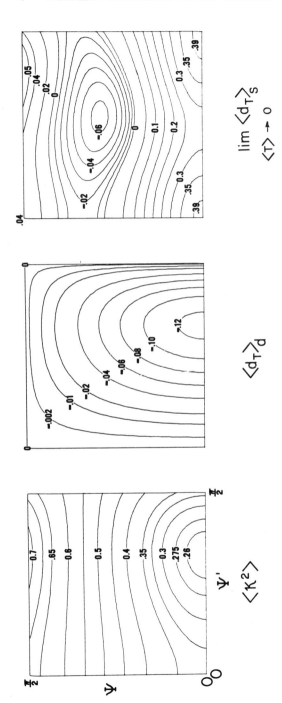

FIG. 24.   Orientation and transfer depolarization factors for Model 5(ii,CC').

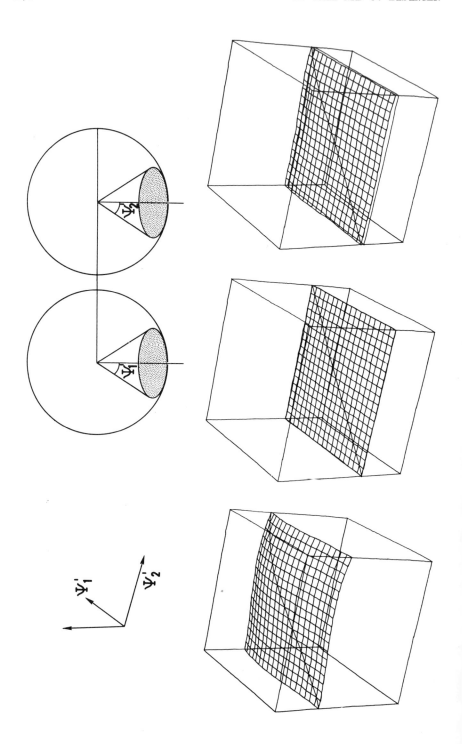

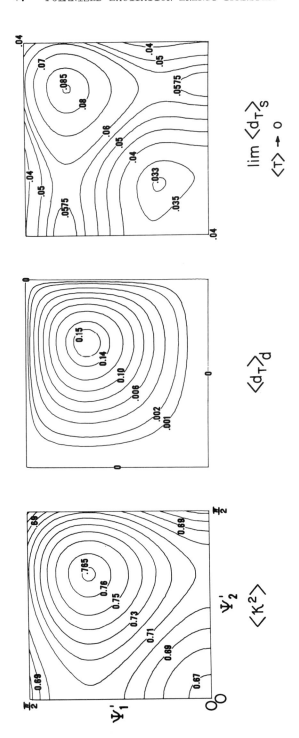

FIG. 25. Orientation and transfer depolarization factors for Model 5(ii,$C'C'$).

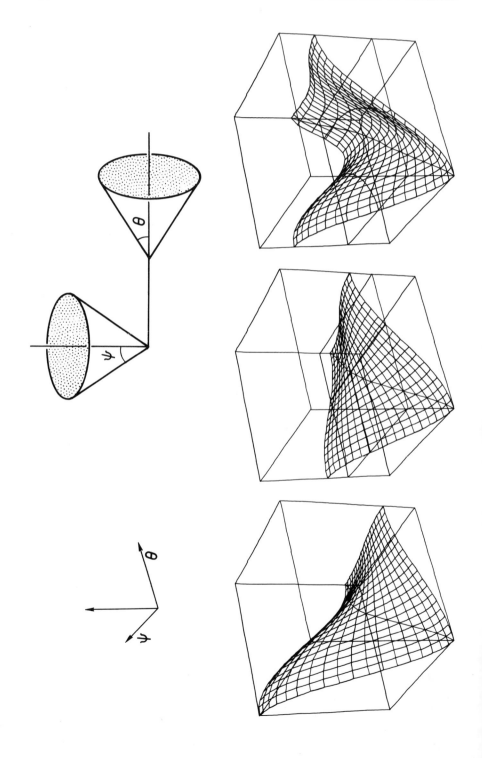

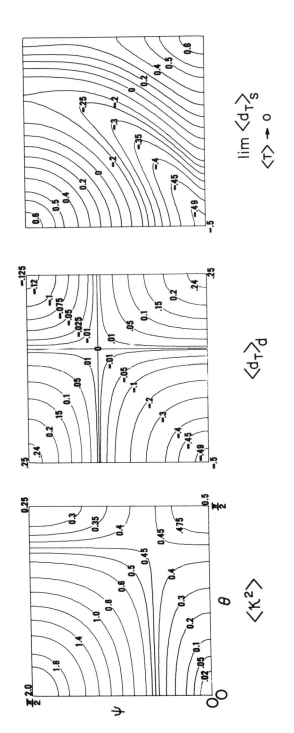

FIG. 26.   Orientation and transfer depolarization factors for Model 5(iii,cc).

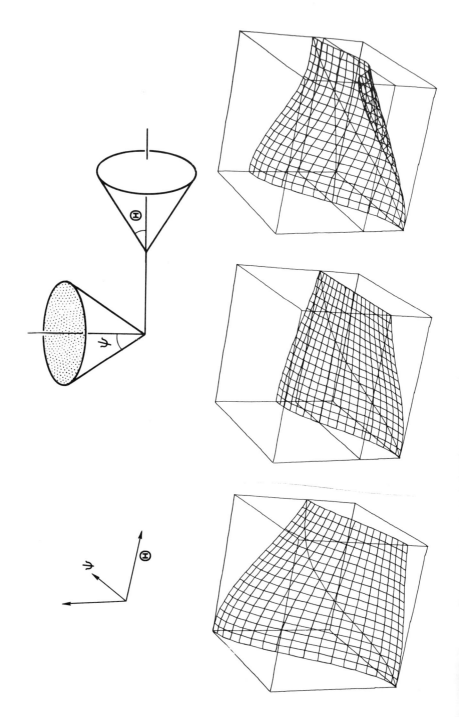

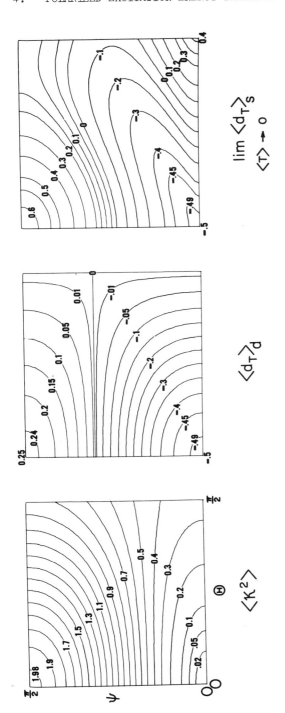

FIG. 27.   Orientation and transfer depolarization factors for Model 5(iii,cC).

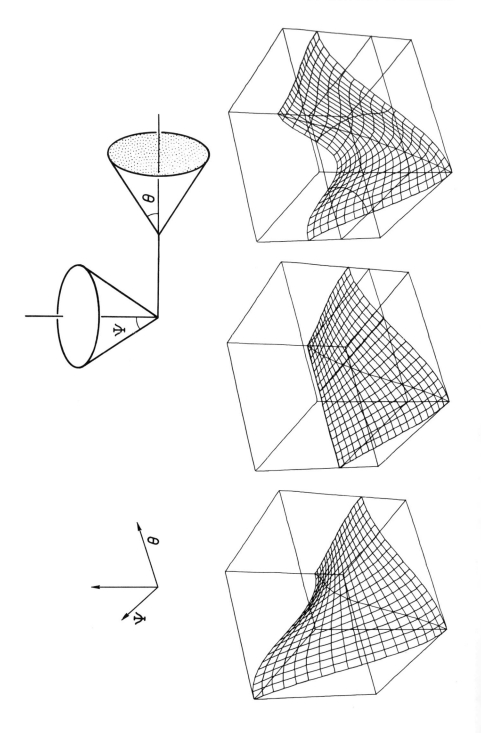

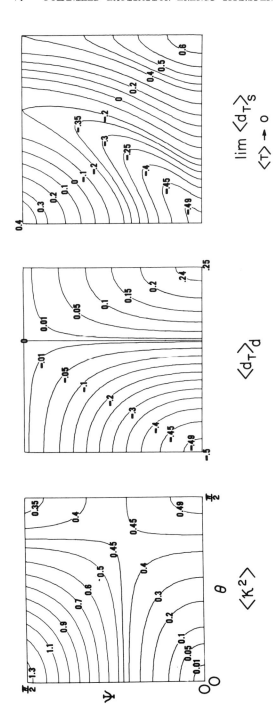

FIG. 28.   Orientation and transfer depolarization factors for Model 5(iii,Cc).

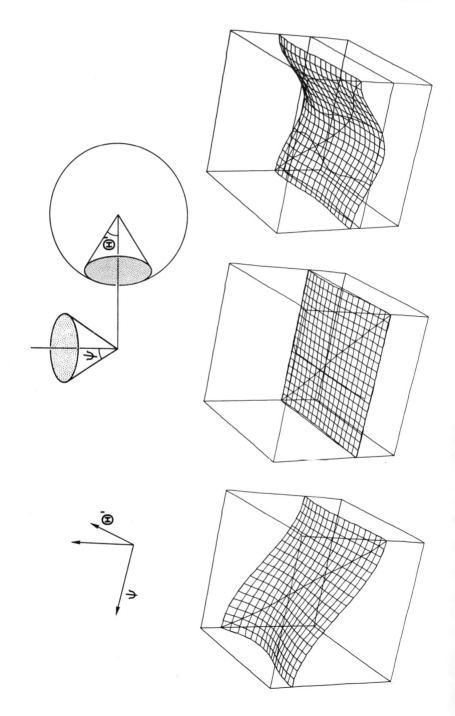

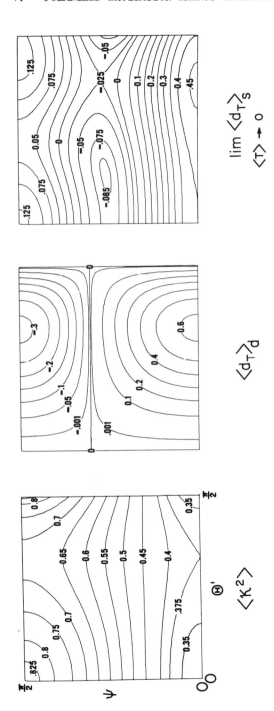

FIG. 29. Orientation and transfer depolarization factors for Model 5(iii,cC').

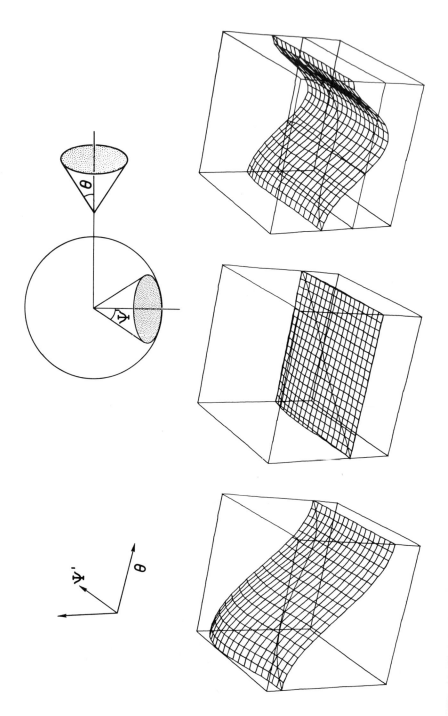

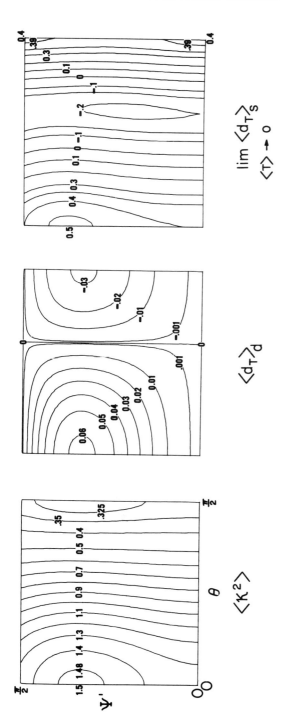

FIG. 30. Orientation and transfer depolarization factors for Model 5(iii,C'c).

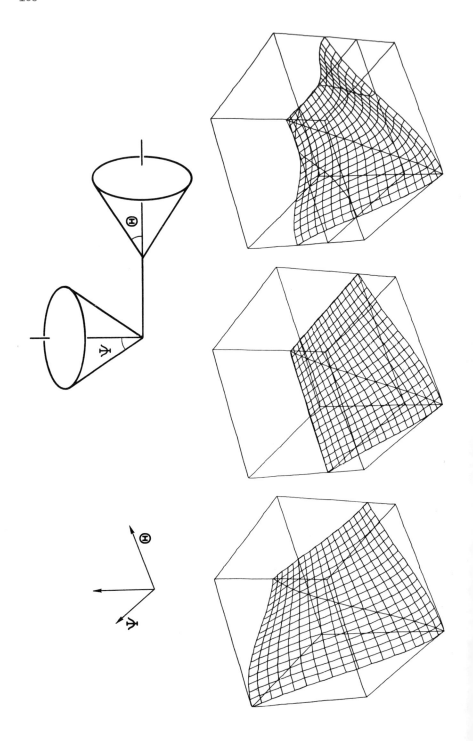

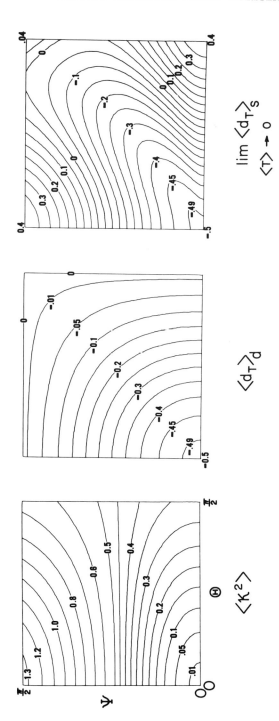

FIG. 31. Orientation and transfer depolarization factors for Model 5(iii,CC).

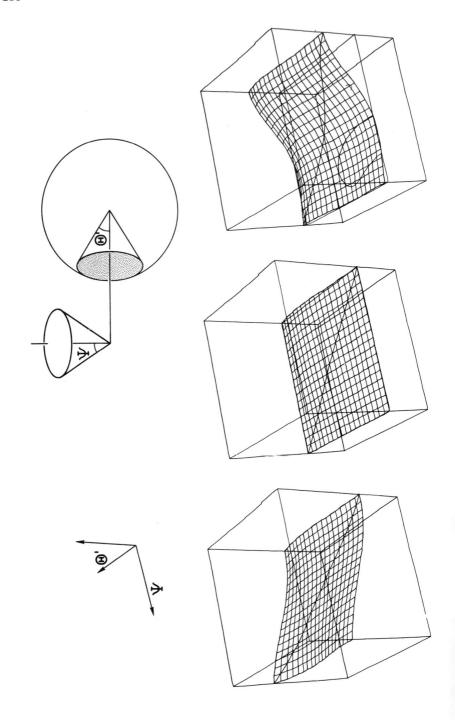

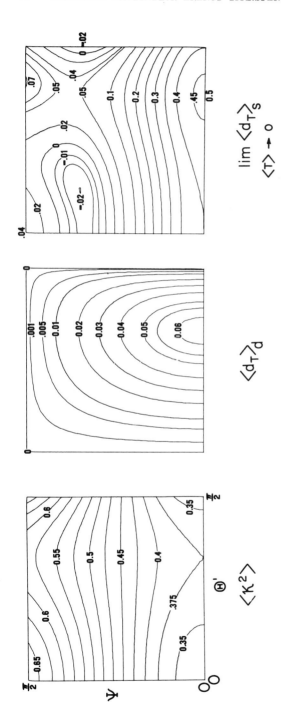

FIG. 32. Orientation and transfer depolarization factors for Model 5(iii,CC′).

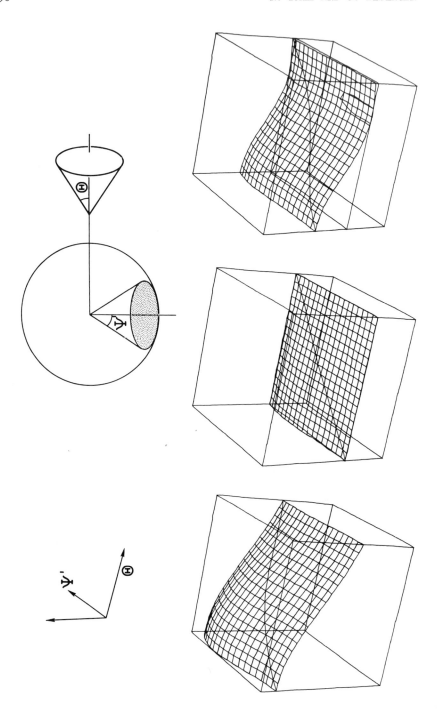

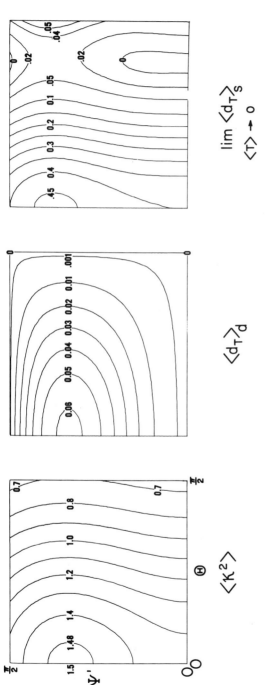

FIG. 33.   Orientation and transfer depolarization factors for Model 5(iii,c′c).

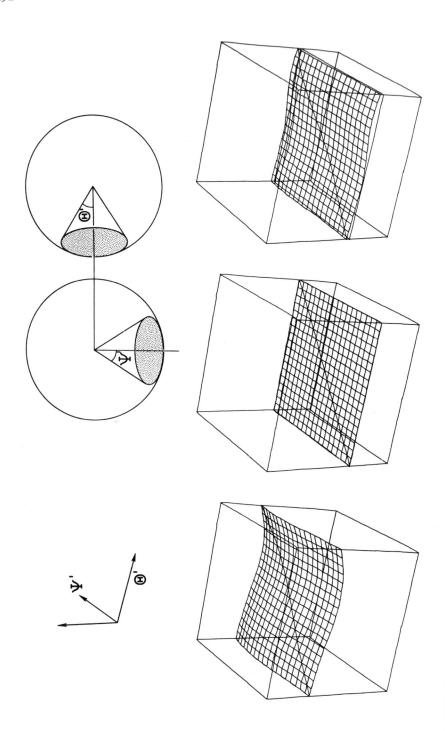

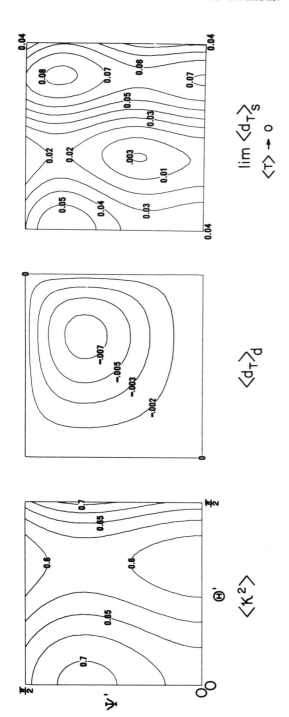

FIG. 34. Orientation and transfer depolarization factors for Model 5(iii,C'C').

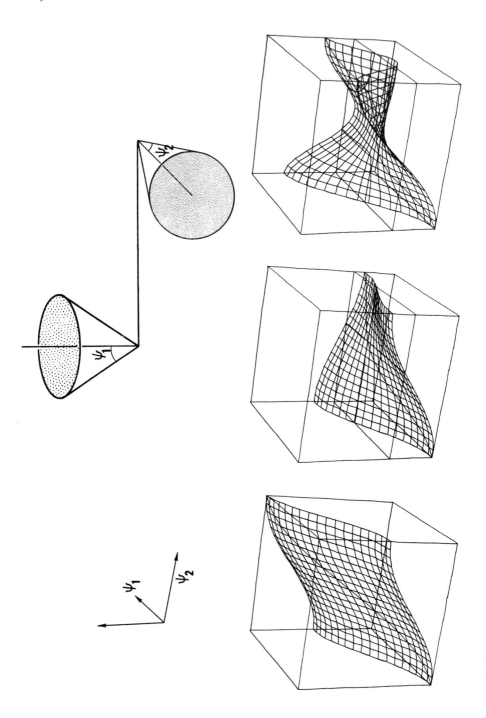

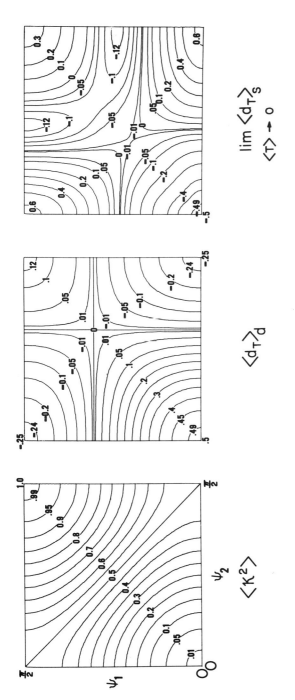

FIG. 35. Orientation and transfer depolarization factors for Model 5(iv,cc).

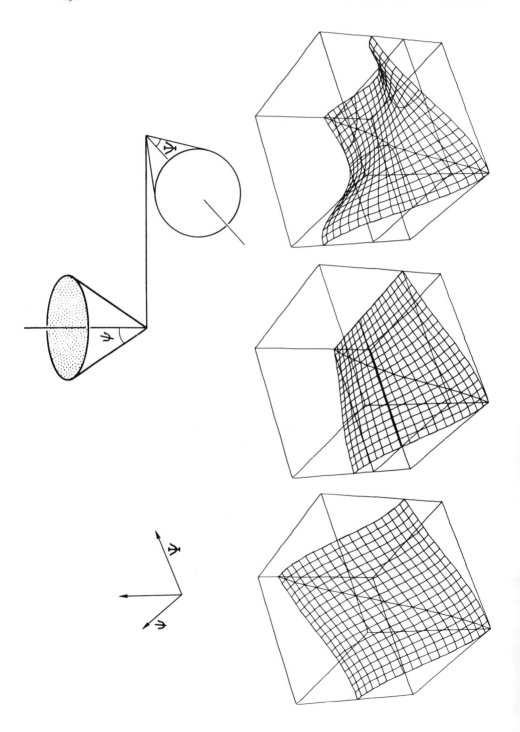

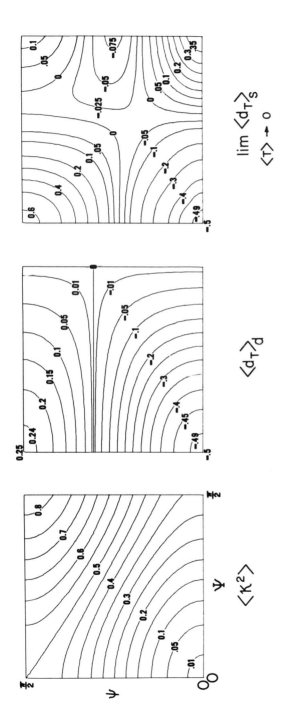

FIG. 36.   Orientation and transfer depolarization factors for Model 5(iv,cC).

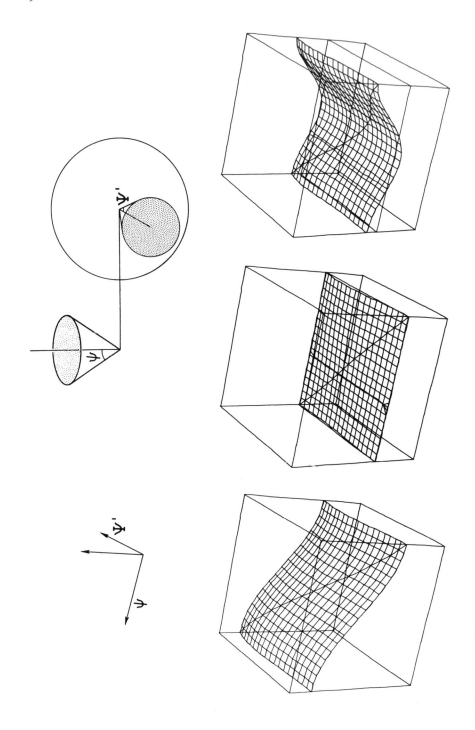

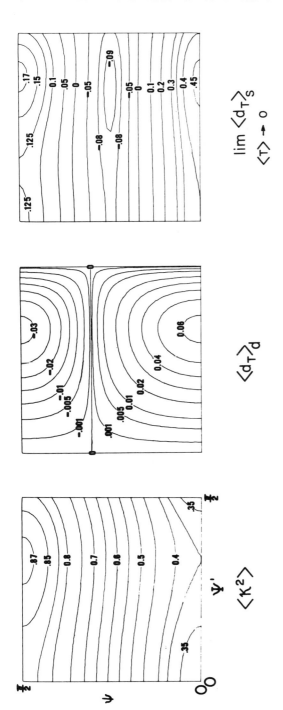

FIG. 37. Orientation and transfer depolarization factors for Model 5(iv,cc').

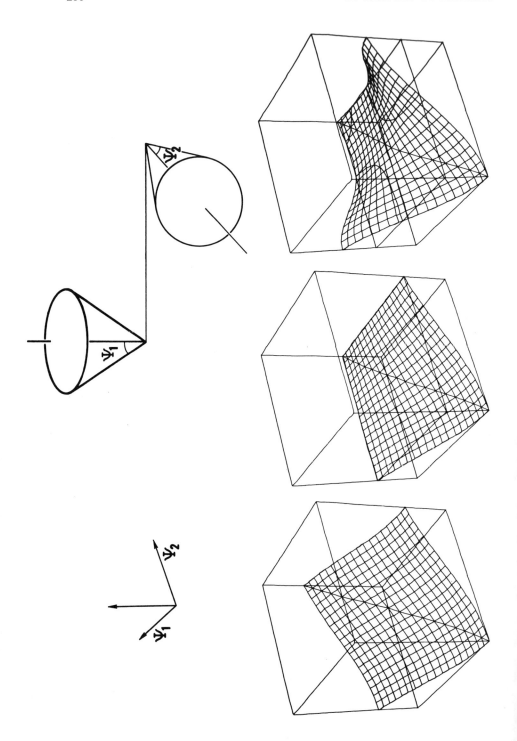

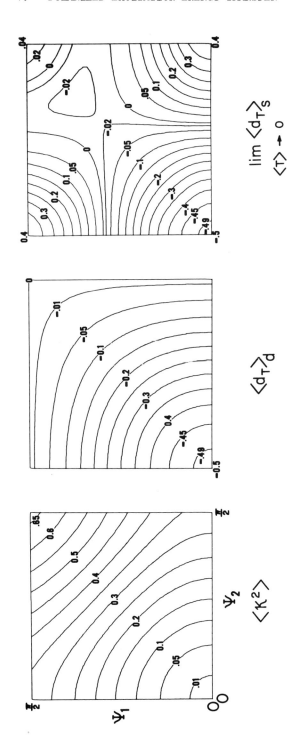

FIG. 38. Orientation and transfer depolarization factors for Model 5(iv,CC).

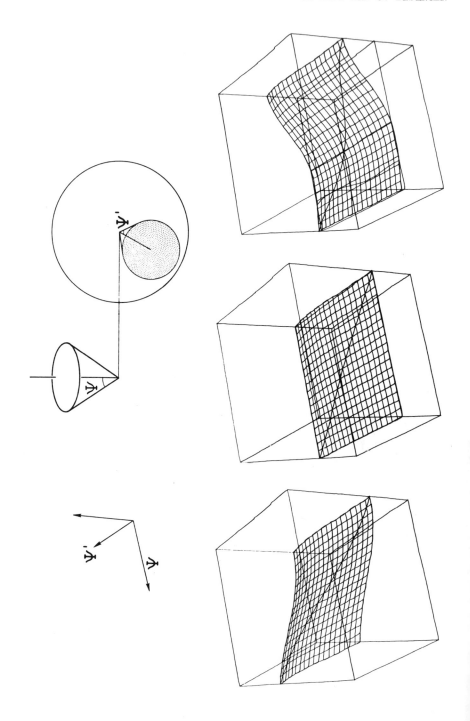

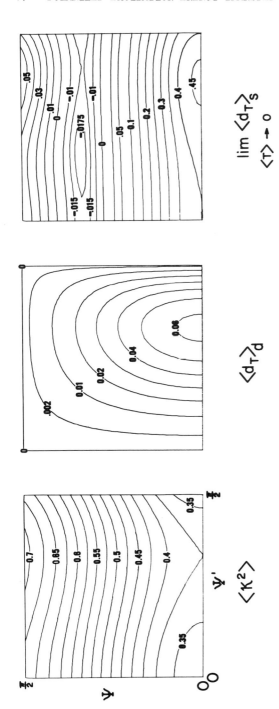

FIG. 39. Orientation and transfer depolarization factors for Model 5(iv,CC').

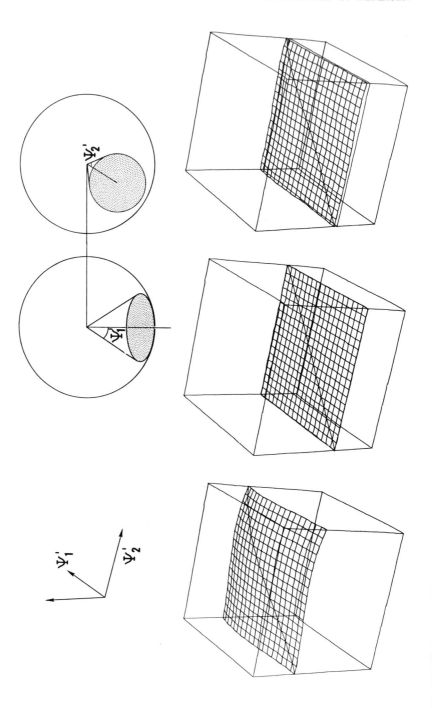

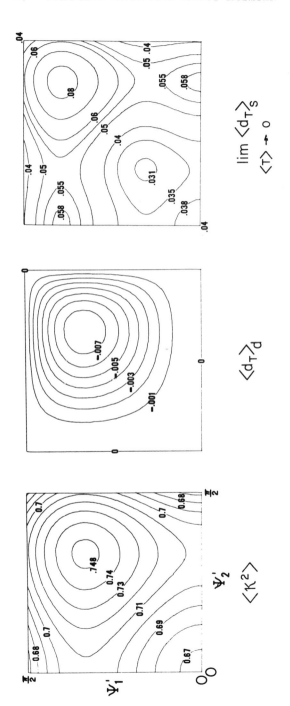

FIG. 40. Orientation and transfer depolarization factors for Model 5(iv,C′C′).

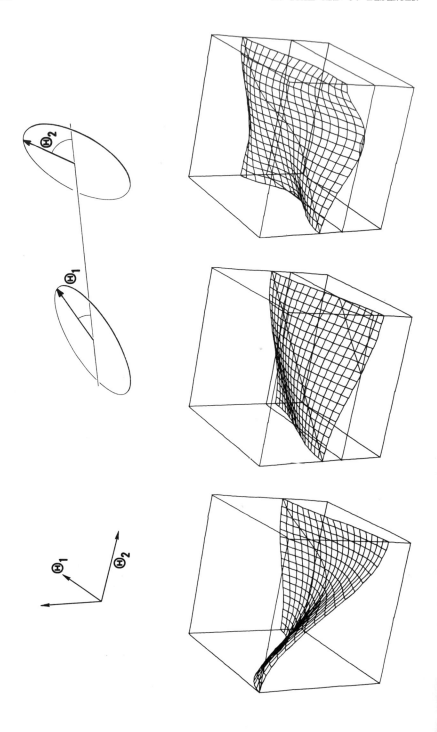

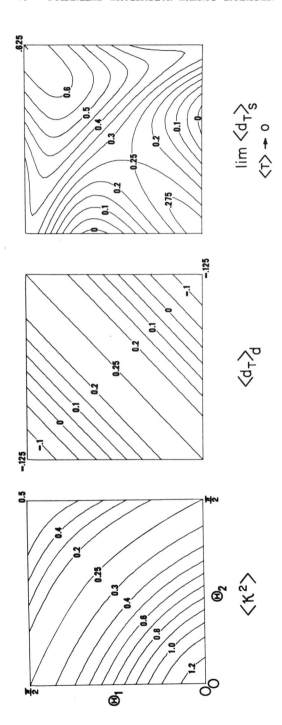

FIG. 41. Orientation and transfer depolarization factors for Model 6(i), $\Phi = 0$.

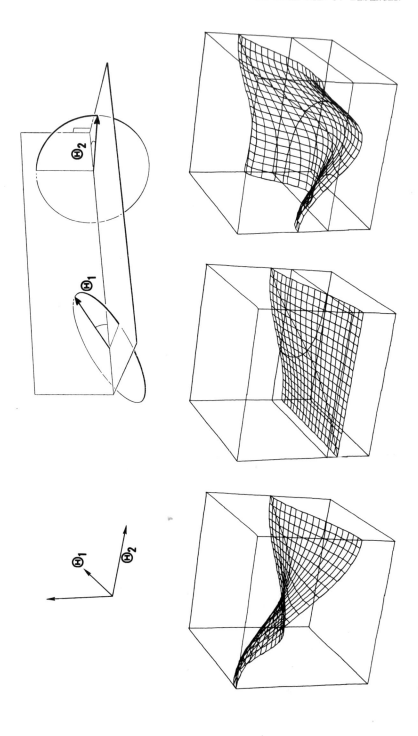

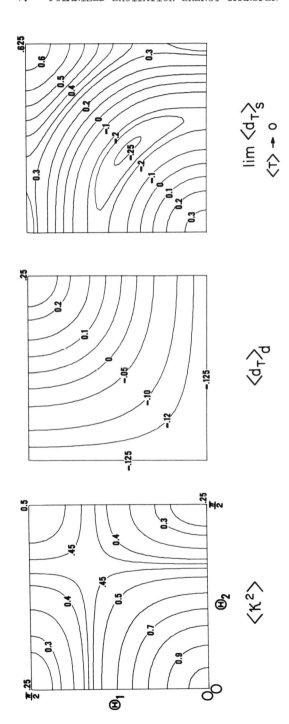

FIG. 42.   Orientation and transfer depolarization factors for Model 6(ii), $\Phi = \pi/2$.

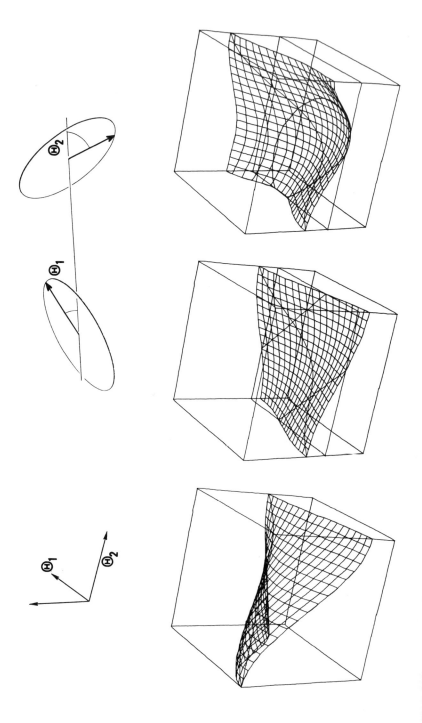

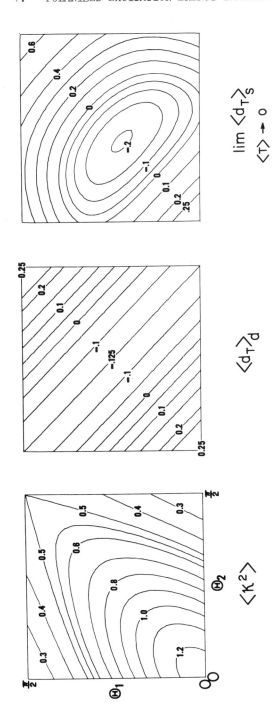

FIG. 43.   Orientation and transfer depolarization factors for Model 6(iii), $\Phi = \pi$.

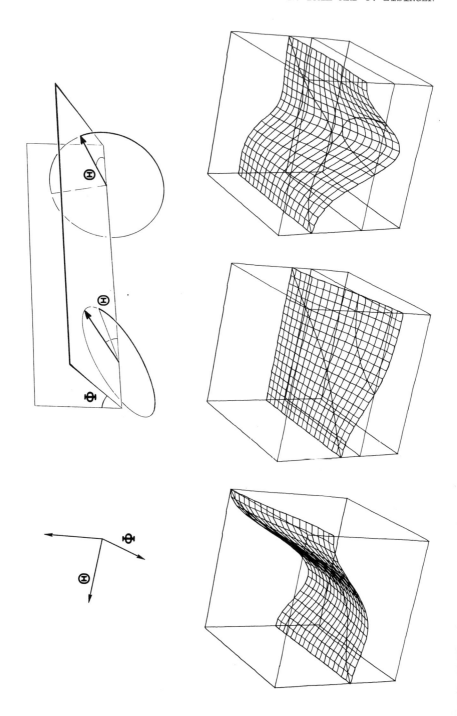

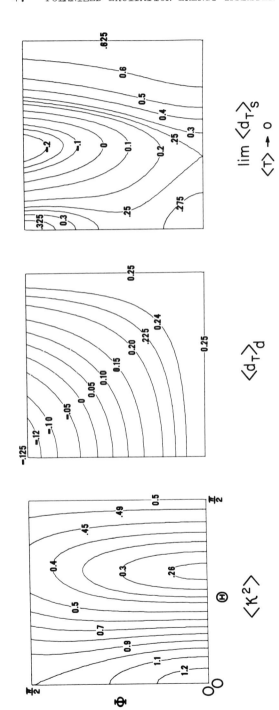

FIG. 44. Orientation and transfer depolarization factors for Model 6(iv), $0 \leq \bar{\Phi} \leq \pi/2$.

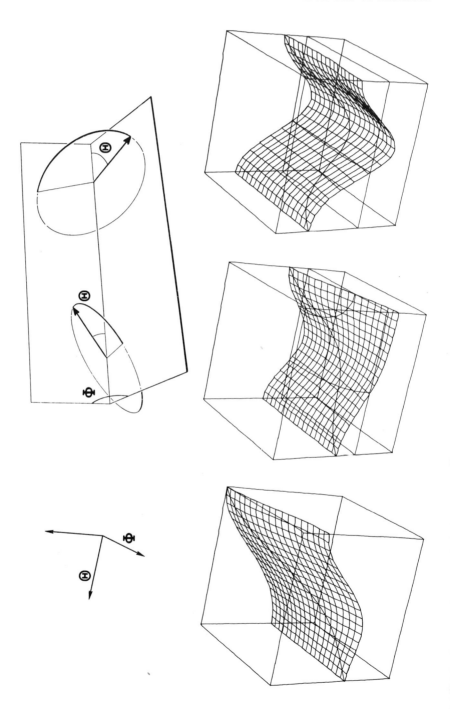

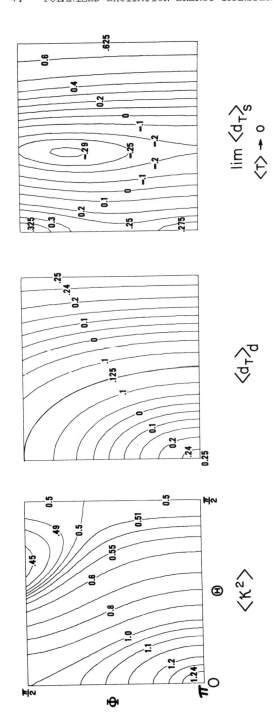

FIG. 45. Orientation and transfer depolarization factors for Model 6(v), $\pi/2 \leq \Phi \leq \pi$.

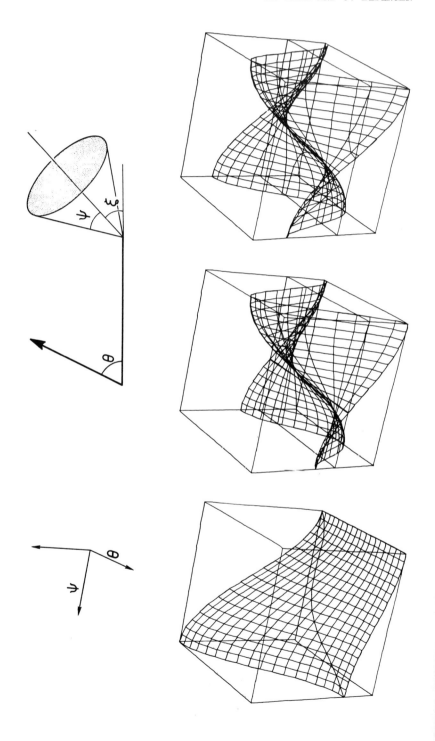

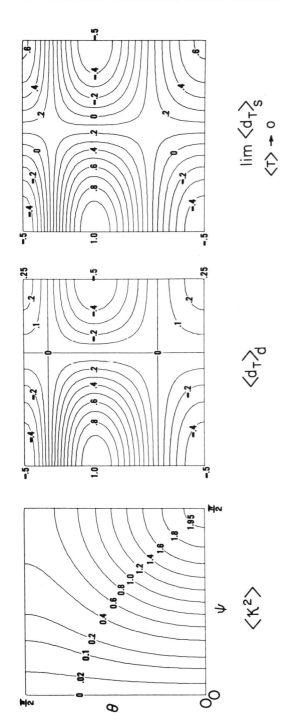

FIG. 46. Orientation and transfer depolarization factors for Model 7(i,c), $\Phi = 0$.

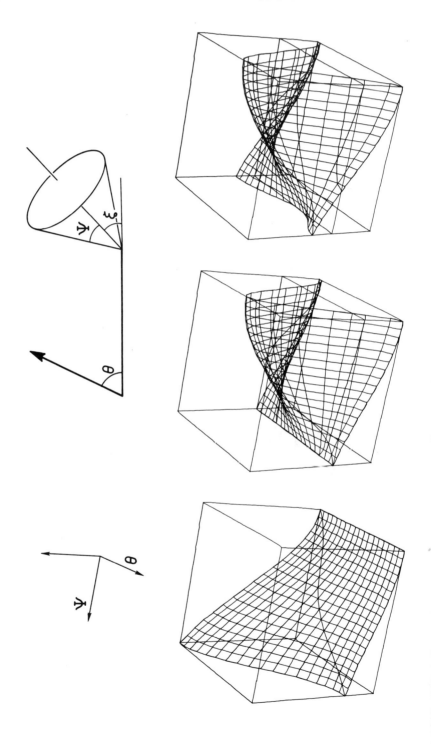

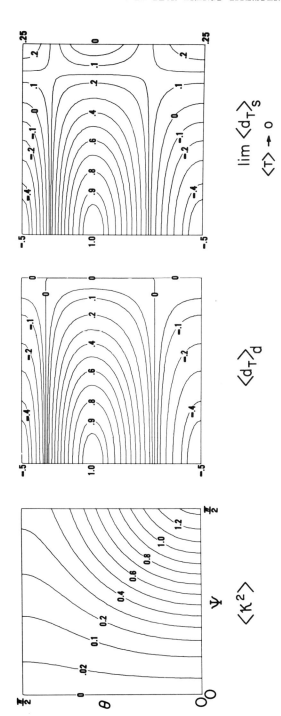

FIG. 47. Orientation and transfer depolarization factors for Model 7(i,C), $\Phi = 0$.

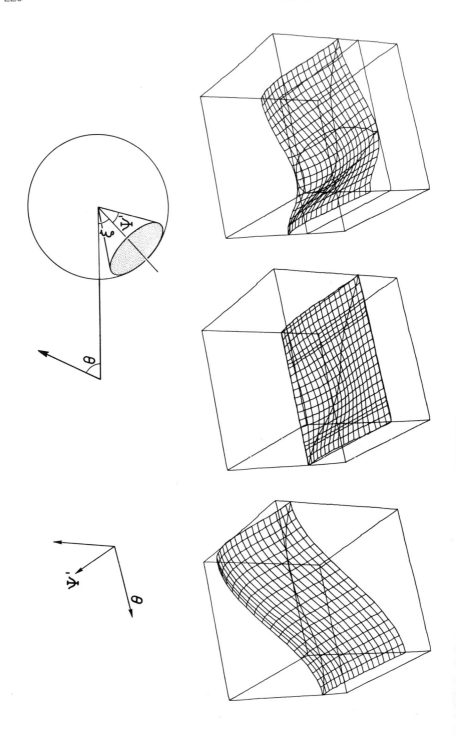

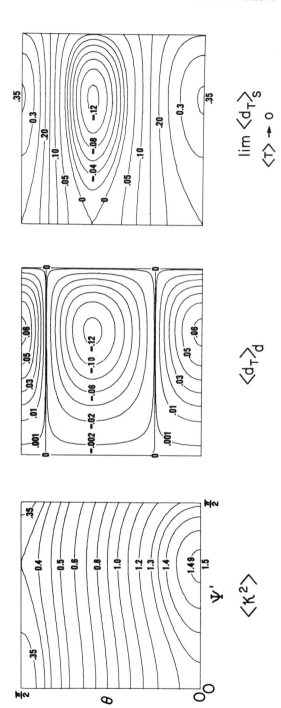

FIG. 48. Orientation and transfer depolarization factors for Model 7(i,c′), Φ = 0.

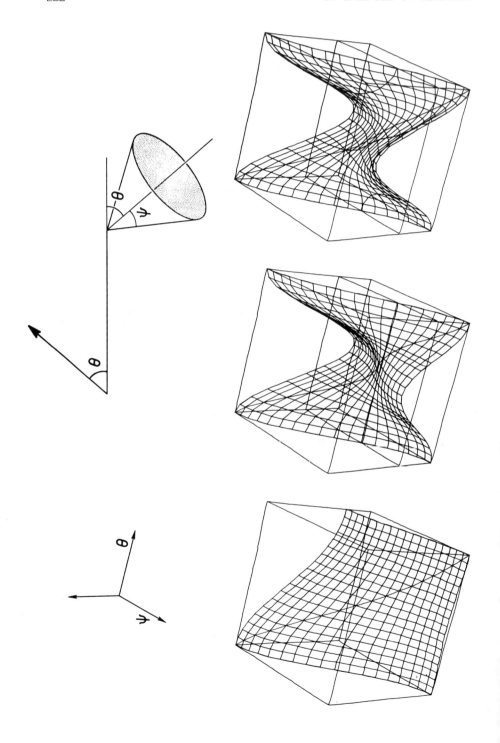

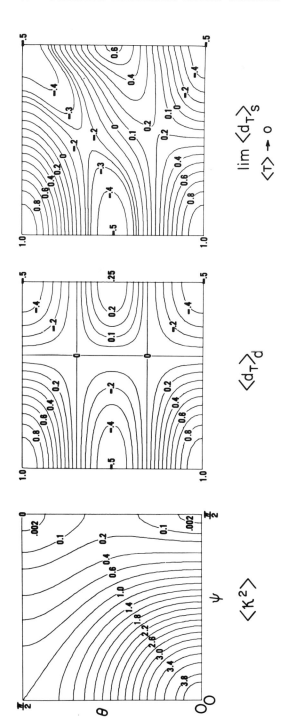

FIG. 49. Orientation and transfer depolarization factors for Model 7(ii,c), $\Phi = \pi$.

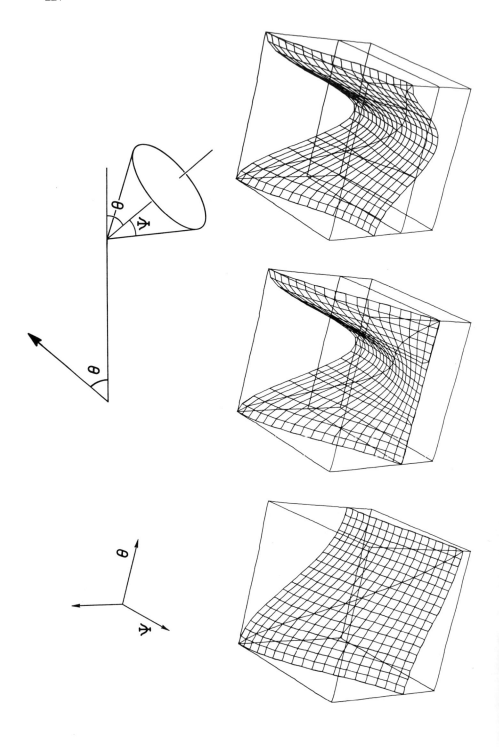

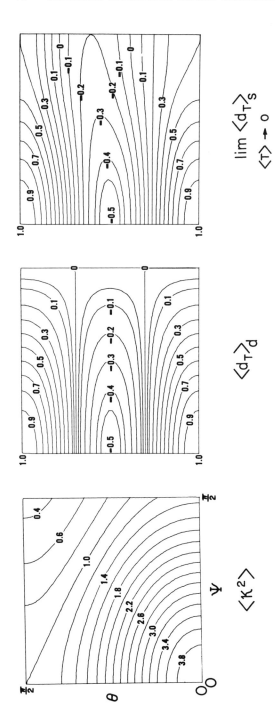

FIG. 50.   Orientation and transfer depolarization factors for Model 7(ii,C), $\Phi = \pi$.

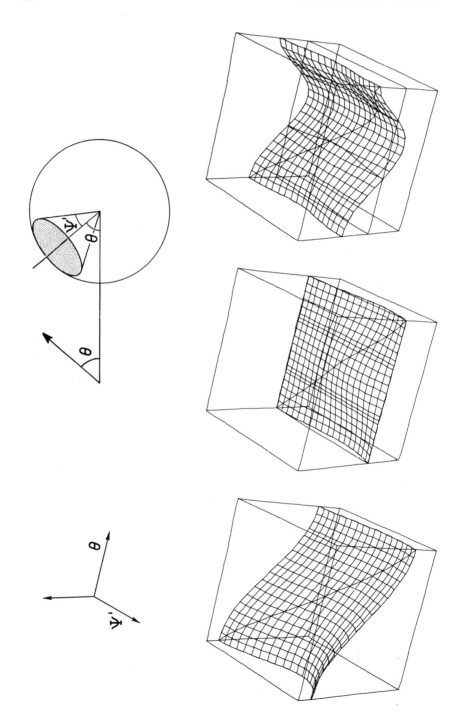

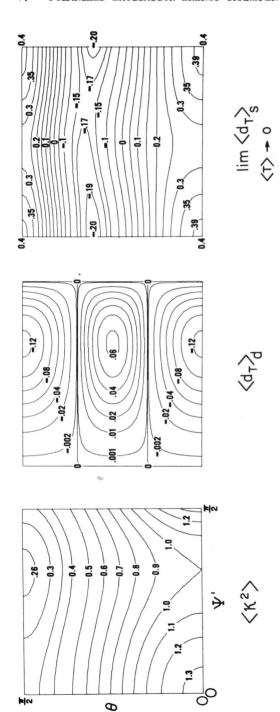

FIG. 51. Orientation and transfer depolarization factors for Model 7(ii,C'), $\Phi = \pi$.

IV. EXPERIMENTAL METHODS

A. Depolarization Factors

In order to utilize fully the information contained in the set
of models described in the previous section, or a more complete set,
it is necessary to estimate the degree of orientational freedom enjoyed by the donor and acceptor separately with respect to the macromolecule. This may be accomplished by measuring the depolarization
of directly excited D and A emission under a dynamic averaging condition, but the depolarization of the luminophore itself must be
separated from that due to Brownian rotation of the macromolecule
as a whole. (In this section only the case of isotropic Brownian
rotation is considered explicitly. Cases of anisotropic rotation
are briefly discussed in Appendix C. Also, should D and A be situated in different parts of a macromolecule joined by a flexible hinge
region, the analysis given here may break down because R may vary
and, furthermore, its variation may be correlated with changes in
the relative angular distributions of D and A, while if the correlation time for the flexibility is of the order of the transfer
time, the averaging is intermediate between the dynamic and static
averaging regimes.) In principle both time resolved and steady state
techniques may be used to determine this depolarization factor, but
it will be seen below that the former lends itself more readily to
meaningful analysis.

1. Decay of Emission Anisotropy

In the first case, the decay of one of the polarized components
of the emission may be measured by using vertically polarized excitation; Jabłonski [33] has shown that for a luminophore which is rigidly
attached to an isotropically rotating substrate, the relative polarized fluorescence decays following a δ-function excitation are given
by

$$I_V(t) = \left(1 + 2r_{om}e^{-t/\tau_m}\right)e^{-t/\tau} \tag{28a}$$

$$I_H(t) = \left(1 - r_{om}e^{-t/\tau_m}\right)e^{-t/\tau} \tag{29a}$$

If natural excitation is used, the decay is given analogously by

$$I_V(t) = \left(1 + r_{onm}e^{-t/\tau_m}\right)e^{-t/\tau} \tag{28b}$$

$$I_H(t) = \left(1 - 2r_{onm}e^{-t/\tau_m}\right)e^{-t/\tau} \tag{29b}$$

where τ_m is the correlation time for isotropic Brownian rotation of the macromolecular substrate and τ the emission lifetime of the luminophore, while r_{om} and r_{onm} are the limiting EA's in the absence of rotation of the macromolecular substrate.

It is convenient to define the time-dependent EA $r(t)$ [33] from Eqs. (28) and (29):

$$r_m(t) = \frac{I_V(t) - I_H(t)}{I_V(t) + 2I_H(t)} = r_{om}e^{-t/\tau_m} \tag{30a}$$

and

$$r_{nm}(t) = \frac{I_V(t) - I_H(t)}{2I_V(t) + I_H(t)} = r_{onm}e^{-t/\tau_m} \tag{30b}$$

Several methods have been used to evaluate $r(t)$ from the experimentally available parameters $i_V(t)$, $i_H(t)$, and $i(t)$ which represent the convolutions of $I_V(t)$, $I_H(t)$, and $I(t) = I_o\exp[-t/\tau]$ with an excitation pulse of finite width. Appropriate sums and differences of $i_V(t)$ and $i_H(t)$ may be analyzed to give $r(t)$ [34,35] [see also Sec. V], or, since all the information is contained in both of them, either one may be analyzed directly [36]. The experimental determination of convoluted time courses of emission and their analyses are discussed in detail in this volume [29,37].

The values of r_{om} and r_{onm} are always smaller (sometimes considerably smaller) than the fundamental values for excitation in the last absorption band, 0.4 and 0.2, respectively, owing to motion of the luminophore with respect to the macromolecule as a whole. However, as has been pointed out by Weber [38], Eqs. (28)-(30) remain valid provided such motion is rapid compared with the macromolecular motion. The limiting values of the EA are then related to the fundamental EA's and the depolarization factor $<d'>_d$ for rapid ("dynamic") restricted movement of the luminophore relative to the macromolecule [Eq. (17)] by

$$\frac{r_{om}}{r_f} = \frac{r_{onm}}{r_{fn}} = \frac{3}{2} < \cos^2 \theta'>_d - \frac{1}{2} = <d'>_d \tag{31}$$

where θ' is the angle through which the transition moment vector of the luminophore moves with respect to the macromolecule during the interval between absorption and emission. In the models of rapid restricted motion over the surface or volume of a cone which are considered in this discussion, $<\cos^2 \theta'>_d$ is a function of the half-angle of the cone. Its values are derived for the surface and volume cases in Appendix C and the depolarization factors determined from them plotted against the half-angle in Fig. 52.

2. Steady State Emission Anisotropy

The depolarization ratio given by Eq. (31) may also be determined from steady state measurements in which the EA is measured either as a function of temperature and viscosity by changing the temperature of the sample, or of viscosity alone at a given temperature by changing the composition of the solvent. Ideally, neither of these changes should have any effect on the structure of the macromolecular complex as a whole or the luminophore itself. Fortunately, this condition is often approached when a temperature range in the order of 0-50° is used or when viscosity over a comparable range is changed by addition of neutral solutes such as sucrose, glucose, mannitol, or glycerol to the aqueous solution which increase

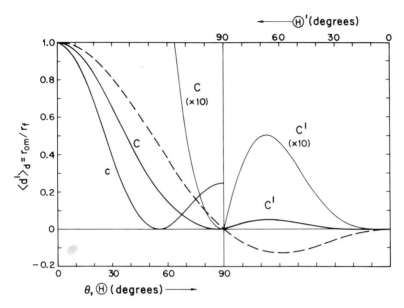

FIG. 52. Dynamic depolarization factors $<d'>_d$ for restricted motion over the surface (c) or volume (C,C′) of a cone. When the motion is over the surface, the half-angle of the cone is θ, when over the volume it is Θ for half-angles less than $\pi/2$ or $\Theta' = (\pi - \Theta)$ for half-angles greater than $\pi/2$ [cf. Eqs. (8a) and (8b)]. The dashed curve represents the depolarization due to motion over the volume of a cone if the initial position of the vector coincides with the cone axis.

its micro viscosity. Solutes which increase only the macro viscosity of the solution, e.g., polyethylene glycol ("Carbowax"), may not be suitable.

The Perrin relationship for isotropic Brownian rotation [38,39] is expressed in terms of EA as

$$\frac{r_{om}}{r} = 1 + \frac{kT\tau}{\eta v} \tag{32}$$

where k is the Boltzmann constant, T the absolute temperature, η the viscosity of the solvent, and v the volume of the (solvated) macromolecule. r_{om} is obtained from the intercept in a plot of r^{-1} against the appropriate variables $T\tau/\eta$ or merely τ/η when keeping the tem-

perature constant as they tend to zero (infinite viscosity, absolute
zero of temperature). The excited state lifetime is included in the
variables, since it is usually temperature dependent for a bound luminophore and may also change on addition of solute. Equation (32)
is obtained from Eq. (30a) by taking the average of $r(t)$ over the
decay time and by equating τ_m^{-1} with the factor $kT/\eta v$.

3. Time Resolution of Segmental Motion

In the idealized situation discussed, the independent luminophore depolarization takes place so fast that its time dependence
cannot be resolved in the nanosecond domain defining the emission
decay time. Similarly, this motion will not be appreciably affected
by changes in viscosity of the solvent in the range indicated above.
With enough time resolution or high enough viscosity, however, or
when, as previously mentioned, the luminophore is attached to a
flexible segment of the macromolecule which may have independent
motion whose correlation time τ_s is only somewhat shorter than that
of the macromolecule as a whole, a second region in the anisotropy
decay curves, e.g. [40,41], or in the Perrin plot, e.g. [42,43],
becomes evident.

No general theory capable of describing the time dependence of
the anisotropy in such cases exists. Gottlieb and Wahl [42] have,
however, solved the case of an isotropically rotating macromolecule
to which a luminophore capable of independent rotation about a single
bond is attached. The decay of the EA is described by three exponential terms, the first of which has the correlation time τ_m of the
macromolecule, while the other two have inverse correlation times
which are linear combinations of τ_m^{-1} and τ_s^{-1}. The pre-exponential
factors may be considered as partial EA's [25], the first one being
r_{om}. A simple derivation of this and equivalent results for cases
of anisotropic rotation are given in Appendix C, along with expressions for the steady state EA which follow immediately from them by

integration over the decay of the emission. Wallach [44] has extended the original result to describe the effect of such rotations
about a series of bonds explicitly.

If τ_s is very much less than a nanosecond, the independent motion may not be detected. Thus, Wahl and coworkers [41] describe
an example where two correlation times are resolved, but the apparent value of the EA at zero time is very low, about half the fundamental, corresponding to a very short third correlation time. As
these workers have pointed out [41], interpretation of reciprocal
Perrin plots is complex unless the two correlation times differ by
about two orders of magnitude or more, even when the decay may be
reasonably approximated by two exponentials [40-43]. Another factor
to be considered in using Perrin plots is the possibility of thermal
activation of the segmental motion [42,43] which necessitates isothermal determination of the course of depolarization as the viscosity alone is changed.

4. Limiting and Fundamental Emission Anisotropy

The value of the time-dependent EA at zero time, $r(0)$, should
ideally coincide with the fundamental EA, r_f, i.e., 0.4 for polarized
excitation in the lowest (nondegenerate) absorption band. Experimentally, $r(0)$ may be obtained by appropriate analysis of polarized
emission decay data corresponding to the high viscosity limit, r_o,
in a Perrin plot. [It may be necessary to use a viscous solution
to obtain $r(0)$ in this way if the segmental relaxation time τ_s is
very short.] However, this limit rarely, if ever, reaches r_f in
practice. Usually r_o turns out to be of the order of 0.32-0.38,
whether the luminophore is bound to a macromolecule or not. The
origin of this discrepancy is not clear. Low apparent anisotropy
values may arise as the result of experimental artifacts such as
imperfect analyzers, misalignment of analyzers or excitation and
observation directions [45], collection of overlarge cones of exciting and emitted radiation [46,47], optical rotation, scattering
or birefringence of the solution [48], or reabsorption of once

emitted light [49], all of which may usually be reduced experimentally until their effect on the measured EA is negligible. It is important to minimize these artifacts in order to be able to obtain a good estimate for the half-angle of the cone over which independent luminophore motion may occur from the experimentally determined value of r_{om}, see Eq. (31) and Appendix C.

Any remaining difference between r_o and 0.4 must be related to the photophysical properties of the luminophore itself in solution and might be explained in several ways. Trivially, the lowest lying absorption band, from which emission takes place, may overlap with higher bands, even to the extent that the higher band absorbs more strongly at the long wavelength edge of the absorption spectrum than the lowest band. This situation gives rise to an anomalous polarization excitation spectrum in which the limiting EA at the long wavelength edge may be smaller than at somewhat shorter wavelengths; tryptophan is a good example of this effect [49]. Usually it is possible to analyze the spectrum into a limiting r_o for the emission transition and another limiting EA corresponding to the higher absorption level in the manner indicated earlier in Sec. II C 2. If there is effectively no such overlap, or if there remains a residual difference between r_o and 0.4 after the appropriate analysis, other mechanisms may be invoked to explain the difference. It may be that the linear oscillator can undergo very rapid torsional vibrations [50] whose magnitude and rate are essentially unaffected in the viscosity range considered. Alternatively, the absorption might be considered as being associated with a two-dimensional [22] or three-dimensional [51] virtual oscillator. On the other hand, the luminophore in either the free state or bound by a single bond to a macromolecular substrate is likely to reside in an extensive solvation shell which, together with the luminophore, forms the appropriate kinetic unit. Motion of the luminophore within the cage would then not be affected by a change in the bulk solvent, except insofar as the structure of the shell itself were altered. Such an explanation is consistent with the finding that while the luminophore fluorescein

free in aqueous glycerolic solution does not have a limiting EA of
0.4 in the Perrin plot, it does after conjugation into polyacryla-
mide during its free-radical polymerization [52,53]; most likely it
is very rigidly bound to a large kinetic unit of the polymer by
several covalent bonds rather than one and thus has no independent
freedom of motion.

The various mechanisms considered above to account for the
differences between the theoretical and observed values of r_o may
be treated, at least to first order, as contributing to the rapid
motion of the luminophore, yielding an overall depolarization factor
$<d'>$ given by Eq. (31) as a function of r_{om}. The value of $<d'>_d$
obtained in this way should still permit estimation of the cone
half-angle from $<\cos^2 \theta'>_d$ to a good approximation, even when the
luminophore is attached to a segment of the macromolecule having a
correlation time of the same order as the emission and transfer
times. Its usefulness in evaluating $<\kappa^2>$ will depend on whether
or not the other luminophore is attached to the same segment of the
macromolecule, a possibility which, in the absence of other information as to the location of the two luminophores, would be strongly
indicated by similar segmental motion correlation times corresponding to a relatively large segment.

B. Transfer Depolarization Factors

The EA $r_{A(D)}$ observed as a result of energy transfer when A
and/or D have some orientational freedom depends on the relative
magnitudes of their independent relaxation rates, the relaxation
rate of the macromolecule, and the transfer rate. The effect of
the macromolecular relaxation may be eliminated by determining the
time or viscosity dependence of $r_{A(D)}$ which may then be analyzed in
a manner which is analogous to that indicated above. Since the
transfer EA observed in a steady state measurement is an average
over the growth and decay of the excited acceptor population, the

Perrin relationship for isotropic macromolecular rotation given in
Eq. (32) becomes

$$\frac{r_{omA(D)}}{r_{A(D)}} = \left(1 + \frac{kT\tau_A}{\eta v}\right)\left(1 + \frac{kT\tau_{D(A)}}{\eta v}\right) \tag{33}$$

where τ_A is the lifetime of acceptor emission and $\tau_{D(A)}$ that of
donor emission in presence of the acceptor (Sec. IV C), $\tau_{D(A)} =$
$\tau_D/[1 + (<R_o^6>/R^6)]$. Equation (33) is equivalent to the analogous
expression derived by Weber [38] for the case in which the emitting
level is populated from a higher level by internal conversion.

Since Eq. (33) is quadratic in $T\tau/\eta$, the corresponding Perrin
plot cannot be extrapolated linearly to yield $r_{omA(D)}$ unless the
quadratic term $\tau_A\tau_{D(A)}/\tau_m^2$ (see Sec. IV A 2) is negligible. When
this condition is fulfilled, as it is for most luminophores used as
macromolecular probes, the intercept of $r_{A(D)}^{-1}$ in the low viscosity,
high temperature region of the Perrin plot gives $r_{omA(D)}$, which is
related to the orientation change θ_T during transfer [see Eq. (16)]
by the expression

$$\frac{r_{omA(D)}}{r_f} = \frac{3}{2}<\cos^2 \theta_T>_d - \frac{1}{2} = <d_T>_d \tag{34}$$

At very high viscosities the observed value of the transfer EA
approaches $r_{oA(D)}$, the value under static averaging conditions, and

$$\frac{r_{oA(D)}}{r_f} = \frac{3}{2}<\cos^2 \theta_T>_s - \frac{1}{2} = <d_T>_s \tag{35}$$

$r_{oA(D)}$ and $r_{omA(D)}$ may be obtained from $r_{A(D)}$ analogously to
r_o and r_{om} by analysis of the polarized decay data. As can be seen
from several of the models presented in Sec. III (Figs. 5-51),
$<d_T>_s$ may be larger or smaller than $<d_T>_d$ or even of opposite
sign.

It is clear from some of the earlier considerations of this
section that very rapid depolarization processes almost always
appear, even in the static limit, and usually correspond to motion

over a relatively small angular range. Analysis of the effect of
this on the static averaging, which may be over a much larger range
of orientations, has not been carried out here, and the values of
$<d_T>_s$ referred to in the models of Sec. III are always the limiting
values in the absence of any such depolarization.

C. Transfer Efficiency

Experimentally, the average transfer efficiencies $<T>_d$ and
$<T>_s$ are obtainable in principle as the limiting values of the
efficiency at zero and infinite viscosity, respectively, although
no relationship equivalent to the linear Perrin relationship for the
EA governs them.

The transfer efficiency may be determined by monitoring either
the change in donor fluorescence when the acceptor is present or by
comparing the absorption spectrum of the D-A pair with the excitation
spectrum of the acceptor emission. The relevant intensities measured
should be proportional to total emission, i.e., to $(I_V + 2I_H)$ for
V-polarized excitation or to $(2I_V + I_H)$ for natural excitation (see
Appendix A). This is most readily accomplished by orienting a polarizer in the observation path at the magic angle $\theta = 54.7$ deg to
V and H, respectively, when the components are passed in the correct
ratios: $\sin^2 \theta = 2 \cos^2 \theta$ [54,55].

The quantum yield of donor emission Φ_D in the absence of an
acceptor is given by

$$\Phi_D = \frac{k_{rD}}{k_{rD} + k_{nrD}} \tag{36}$$

where k_{rD} is the rate constant for emission and k_{nrD} the rate of
radiationless deactivation of S_1. $(k_{rD} + k_{nrD})$ is equal to τ_D^{-1}, the
reciprocal of the donor excited state lifetime. The quantum yield
of the donor in the presence of an acceptor to which singlet excited

energy is transferred at a rate k_T given in Eq. (1) is expressed, for a D-A pair i, by

$$\Phi_{DAi} = \frac{k_{rD}}{\tau_D^{-1} + k_{Ti}} \tag{37}$$

By combining these yield expressions and by substituting into Eq. (6), it can be seen that the average transfer efficiency is related to the donor quantum yield in presence and absence of the acceptor by

$$<T> = 1 - \frac{<\Phi_{D(A)}>}{\Phi_D} \tag{38}$$

where the averages are taken over all configurations. In the dynamic averaging limit (also if all D-A pairs have the same fixed relative orientations), the average value of $\Phi_{D(A)}$ is the same for each D-A pair, and Eq. (38) may be replaced by

$$<T>_d = 1 - \frac{\tau_{D(A)}}{\tau_D} \tag{39}$$

where $\tau_{D(A)}$ is the donor lifetime in the presence of the acceptor. Under static averaging conditions, on the other hand, the ensemble of D-A pairs cannot be characterized by a single exponential decay time. For each configuration i, the lifetime will be weighted by the yield [13,56] so that the equivalent of Eq. (39) may not be used to relate $<T>_s$ to the change in donor lifetime.

Alternatively, observation of the excitation spectrum of acceptor emission may be used to obtain an estimate of the transfer efficiency. If a fraction $a_D(\lambda)$ of excitation light of wavelength λ absorbed by the D-A pair is absorbed by the donor, then the overall emission quantum yield of the acceptor $\Phi(\lambda)$ is given by

$$\Phi(\lambda) = [1 - a_D(\lambda)]\Phi_A + a_D(\lambda)<T>\Phi_A \tag{40}$$

where Φ_A is the quantum yield of directly excited acceptor emission (which is independent of the excitation wavelength) and $<T>$ the average transfer efficiency from D to A. If $A_A(\lambda)$ and $A_D(\lambda)$ are the absorbances of A and D at the excitation wavelength, $a_D(\lambda)$ is given by

$$a_D(\lambda) = \frac{A_D(\lambda)}{A_D(\lambda) + A_A(\lambda)} \tag{41}$$

and if $I(\lambda)$ and $I_A(\lambda)$ are the intensities of acceptor emission (in arbitrary units) in the presence and absence of D absorption, then

$$\frac{\Phi(\lambda)}{\Phi_A} = \frac{I(\lambda)/[A_D(\lambda) + A_A(\lambda)]}{I_A(\lambda)/A_A(\lambda)} \tag{42}$$

Strictly speaking, a_D is related to the fractions of light absorbed by D and A rather than to their absorbances. However, as long as the absorbance of the sample is small, 0.02 or less, as is usually required experimentally to minimize trivial reabsorption, e.g. [9,57,58], a_D is linearly related to the fractional absorption, and Eq. (41) represents an excellent approximation.

By substituting Eqs. (41) and (42) into Eq. (40) and by solving for the transfer efficiency,

$$<T> = \frac{A_A(\lambda)}{A_D(\lambda)} \left(\frac{I(\lambda)}{I_A(\lambda)} - 1 \right) \tag{43}$$

which, on introducing the normalization $A_A(\lambda) = I_A(\lambda)$, reduces to the simpler expression

$$<T> = \frac{I(\lambda) - A_A(\lambda)}{A_D(\lambda)} \tag{44}$$

with which $<T>$ is readily determined by comparing the absorption spectra of the D-A pair and A alone with the excitation spectrum of emission normalized to them in the region where A alone absorbs.

V.   ILLUSTRATIONS OF DISTANCE DETERMINATIONS

The preceding sections summarize how intramolecular distances (R) between donor and acceptor chromophores may in principle be

determined from a knowledge of the efficiency of energy transfer between them and a characteristic distance R_o which is in turn a function of their mutual orientation. It has also been indicated how
ranges of possible values of the orientation factor κ^2 may be estimated from a knowledge of the donor and acceptor depolarization factors ($<d'_D>_d$ and $<d'_A>_d$) along with the transfer efficiencies ($<T>_d$
and $<T>_s$) and transfer depolarization factors ($<d_T>_d$ and $<d_T>_s$).

   Ideally, polarized energy transfer experiments should include
determinations of all of the aforementioned six parameters. In
practice, the values of $<d'_D>_d$, $<d'_A>_d$, $<T>_d$, and $<d_T>_d$ are the
most important and, experimentally, the most accessible quantities.
Thus, Fig. 52 permits estimation of the motional freedom of the
donor and acceptor, once $<d'_D>_d$ and $<d'_A>_d$ have been measured.
These estimates determine particular values of $<d_T>_d$ for any given
model describing the relative orientation of $\underline{D}$ and $\underline{A}$, which may then
be compared with an experimentally determined value of $<d_T>_d$. For
each model which is consistent with the experimental values of these
three dynamic depolarization factors, a possible value of $<\kappa^2>$ is
obtained.

   The number of models will be further reduced if appropriate
measurements in the static limit can be made. It must be borne in
mind, however, that $<d_T>_s$ depends on $<T>_s$ and that the contour
maps offered in Sec. III give values for $<d_T>_s$ in the limit of zero
transfer efficiency only. For finite transfer efficiencies, the
relationship between $<T>_s$ and $<d_T>_s$ for a given model may be obtained from Eqs. (10) and (24), permitting the elimination of models
for which consistency between $<T>_s$ and $<d_T>_s$ cannot be achieved.
$<d_T>_s$ always lies between $<d_T>_d$ and $\lim_{<T>\to 0} <d_T>_s$ (see Sec. III).

   While no such complete polarized energy transfer experiments
have been reported in the literature to date, the example below
illustrates how even a knowledge of $<d'_A>_d$ and $<d'_D>_d$ alone permits
a systematic analysis of unpolarized energy transfer experiments.
(Some transfer depolarization studies have in fact been carried out

on a number of small model compounds which exhibit essentially 100% transfer efficiency [59]. Analysis of such cases differs somewhat from the above and will be presented elsewhere.)

A. The Structure of Transfer RNA

The phenylalanine-specific tRNA from bakers' yeast ($tRNA^{phe}$) contains a fluorescent base (Y) adjacent to its anticodon. Beardsley and Cantor [3] attached three different acridine dyes covalently to the CCA terminus of this molecule and measured the efficiency of energy transfer from the Y base to them. From their measurements the authors calculated the distance between the Y base and the dye to be between 45 and 59 Å, assuming the dynamic random average value of 2/3 for κ^2. On the other hand, they indicated that the dyes are probably immobilized with respect to the macromolecule because of intercalation (or perhaps stacking), while the Y base has some orientational freedom. Their studies of the time-dependent depolarization of fluorescence of the Y base [60] indicated a rotational correlation time of 10 nsec, while those on an intercalated ethidium bromide probe gave a time of 25 nsec [61]. They concluded from the second result that the tRNA molecule in solution is slightly prolate (axial ratio ~ 2-2.5) and from the first that the Y base, or a segment to which it is attached, has some independent flexibility.

The present authors have recently remeasured the polarized decays of the Y base fluorescence of this tRNA under identical conditions [62]. In order to extract the EA decay from the polarized intensity data, $i_V(t)$ and $i_H(t)$, it is necessary to normalize them. This was accomplished by obtaining an independent value for the steady state EA, $<r> = 0.185 \pm 0.002$, and by comparing it with the integrated values of $i_V(t)$ and $i_H(t)$:

$$<r> = \frac{\int i_V(t) \, dt - \alpha \int i_H(t) \, dt}{\int i_V(t) \, dt + 2\alpha \int i_H(t) \, dt} \tag{45}$$

where α is the normalization factor. After Eq. (45) was solved for α, the data yielded a single exponential rotational correlation time of 20 ± 2 nsec and a limiting EA value $r_{om} = 0.241 \mp 0.012$, obtained by use of the relationship

$$\frac{<r>}{r_{om}} = \frac{\tau_m}{\tau + \tau_m} \tag{46}$$

which represents Eq. (30) averaged over the single exponential decay of emission $I(t) = I_o e^{-t/\tau}$ [33]:

$$<r> = \frac{\int_0^\infty I(t)r(t)\, dt}{\int_0^\infty I(t)\, dt} = r_{om} \frac{1}{\tau} \int_0^\infty e^{-t(1/\tau + 1/\tau_m)}\, dt \tag{47}$$

If the emission and/or EA are functions of more than a single exponential decay, the analogous expression for the average EA is

$$<r> = \frac{\int_0^\infty \Sigma_i I_i(t)r_i(t)\, dt}{\int_0^\infty \Sigma_i I_i(t)\, dt} \tag{48}$$

τ, the single exponential decay time of total emission, was 6.0 ± 0.1 nsec, in good agreement with Beardsley, Tao, and Cantor's value of 6.3 nsec [60]. An intensity proportional to total emission was measured via a polarizer at the magic angle (see Sec. IV C). It was shown that the data could also be fitted with these authors' value of 10 nsec for the rotational correlation time by using an incorrect normalization of $i_V(t)$ and $i_H(t)$ but that this leads to an r_{om} value which is incompatible with the measured steady state EA value, $<r>$.

It is worthwhile stressing that the normalization procedure based on $<r>$ is to be preferred over the other procedures indicated [see Sec. IV A 1], since the determination of a steady state EA

value is at present about an order of magnitude more precise than
that obtained in fitting polarized decay curves.

The comparatively low value of r_{om} obtained above indicates
rapid restricted motion of the Y base, the average value corresponding to cone half-angles of about 23 and 32.5 deg for the cone surface
and volume models, respectively (see Appendix C and Fig. 52). Similarly, for acriflavine covalently attached to the 3'-terminus of
unfractionated tRNA, the value $r_{om} = 0.308$ corresponding to the
limiting value of p_n obtained by Churchich [63] from a Perrin plot
indicates a rapid restricted motion of the acceptor group too, corresponding to cone half-angles of about 16.5 and 23.5 deg.

In the absence of additional information, in particular about
the depolarization during energy transfer, the most suitable model
for Beardsley and Cantor's experiment is thus one in which both
donor and acceptor move rapidly during the transfer time over restricted regions, the axes of donor and acceptor motions having completely unknown orientations with respect to each other and to the
separation vector. In order to gauge the uncertainty in R introduced by ignorance of this relative orientation, those models among
the ones discussed in Sec. III which lead to maximum and minimum
values of $<\kappa^2>$ (and hence of R) are examined in Table 1 and compared with those in which the acceptor is rigid while rapid motion
of the donor is either restricted or random.

It is evident that R may take on a much wider range of values
(34-76 Å) than had been supposed. It is interesting to note that
the upper limit is close to the separation of approximately 80 Å
between the CCA terminus and the anticodon which has recently been
determined from the crystallographic studies of Kim and coworkers
[64], while the lower limit is not inconsistent with the separation
of about 30 Å suggested in the structure proposed by Jang [65].

Table 1 also shows that for the relatively small cone half-
angles involved in this case, there is very little difference between

TABLE 1

Models Leading to Maximum and Minimum Values of $\langle \kappa^2 \rangle$

Model (Appendix B)		Figure	Y——A	κ^2	$R/R_{2/3}$	R_{max} (Å)	R_{min} (Å)
-		-		0.667	1.00	59	45
1	$(\theta = 0)$	5		1.333	1.12	66	-
4(i)	c $(\theta = 0)$	8		3.389	1.31	77	
	C	9		3.406	1.31		
5(i)	cc	8		3.122	1.29	76	-
	cC	9		3.122	1.29		-
	Cc	9		3.137	1.29		-
	CC	17		3.138	1.29		-
1	$(\theta = \frac{\pi}{2})$	5		0.333	0.89	-	40
4(iii)	c $(\theta - \frac{\pi}{2})$	13		0.0763	0.697	-	29
	C	14		0.0742	0.697	-	
5(iv)	cc	35		0.117	0.748	-	34
	cC	36		0.117	0.748	-	
	Cc	36		0.115	0.746	-	
	CC	38		0.115	0.746	-	

the average orientation factors derived from the surface and volume distribution models for restricted motion of the luminophores, indicating that the value of $<\kappa^2>$ is rather insensitive to the precise model used to describe this motion.

B. Distribution of Donor-Acceptor Separations

It has been suggested that by labeling the ends of flexible polymers with suitable pairs of donors and acceptors it is possible to determine the distribution of their end-to-end distances. Cantor and Pechukas' method [66] requires steady state measurements of the transfer efficiency for several D-A pairs characterized by different R_o values, while Steinberg and co-workers [67] have suggested analyzing the decay kinetics of the donor for any particular D-A pair. Both methods assume an average value for κ^2 which is independent of the polymer conformation. This assumption is valid only if both luminophores have complete orientational freedom with respect to the polymer, or if changes in end-to-end distance do not affect the relative D-A orientations. However, it is much more likely that the luminophores have a restricted motional freedom and that flexing of the polymer leads to changes in their relative orientations which correspond to wide ranges of average κ^2 values, invalidating such analyses.

VI. CONCLUDING REMARKS

In the preceding sections, techniques for analyzing polarized energy transfer experiments have been outlined and illustrated. Section V shows, moreover, that even if the depolarization of transferred energy is unknown or cannot be measured (e.g., if the acceptor is nonfluorescent), it is of great value to determine the depolarization of the donor and, where possible, acceptor independently, since this permits the systematic estimation of upper and lower limits of κ^2. In the present chapter an analysis is offered for a number of simple models, although a more general solution of this problem appears feasible in principle. The models presented here should suffice, however, to demonstrate the intimate connection between depolarization and the dipolar orientation factor, and hence between the former and an interchromophore separation as determined by an energy transfer experiment.

Mention should also be made of the use of depolarization as an empirical probe of conformational changes. Such applications have to date been confined to cases in which the depolarization of a single probe is monitored. It is clear, however, that the measurement of the transfer depolarization may be a particularly sensitive tool for detecting conformational changes which affect the relative orientations of two luminophores at different locations in the macromolecule as well as local conformational changes affecting the motion of the luminophores separately.

ACKNOWLEDGMENTS

This work was made possible by B. C. Chambers who provided the numerical integrations for the models discussed in Sec. III and by W. E. Blumberg who wrote the interactive graphics program for plotting the three-dimensional figures and contour maps and to whom the authors are grateful for many useful discussions. The authors also wish to thank Ewa A. Dale and Marion Riley for their invaluable help in preparing the manuscript.

APPENDIX A: POLARIZED EMISSION SPECTROSCOPY

The fundamental relationships of polarized emission spectroscopy
will be derived in a general coordinate system and will be restated
for the laboratory coordinate systems in common use. While these
results will be formulated in terms of emission anisotropy [23,24],
equivalent relationships in terms of the degree of polarization will
also be given.

1. Emission Anisotropy of
Isotropic Solutions of Luminophores

Consider a linearly polarized collimated beam of light whose
electric vector and propagation direction define the z- and y-axes
of an Oxyz coordinate system. Assume that some of the light incident on a sample at the origin O is absorbed by the sample and then
emitted. The analysis of polarized emission presented here applies
only to systems of incoherently emitting luminophores. Special
considerations obtain in cases of coherent (laser) emission. In
general, the emission observed in an arbitrary direction will be
polarized to a greater or lesser extent depending upon this direction, the nature of the sample, and the change in orientation, if
any, between absorption and emission vector directions.

The emission from a single luminophore is fully polarized in a
plane containing the transition dipole and the propagation direction.
Furthermore, the probability of emission in a particular propagation
direction is proportional to the square of the sine of the angle
between the transition dipole moment vector $\underline{E}$ and that direction.
If all the transition dipoles in a sample were parallel to each
other and to a particular direction, the maximum emission intensity
would be polarized in this direction and observed in the plane normal to it. The minimum intensity of zero would be observed along
the transition dipole direction. The emission anisotropy (EA) is

defined with respect to the Oxyz coordinate system described above
by

$$r_p = \frac{I_z - I_y}{I} \qquad\qquad (A1)$$

where I is the sum of any three orthogonally polarized intensities,
e.g. $(I_z + I_y + I_x)$, and the subscript p denotes linearly polarized
excitation.

In order to relate the EA to the emission transition moment
vector, it is useful to restate Eq. (A1) in terms of angular coordinates. Consider an arbitrary emission dipole moment vector $\underline{E}$
oriented at Euler angles ξ, ζ, and θ with respect to the x, y, and
z axes of the excitation-based coordinate system, while α, β, and φ
are the associated azimuthal angles in the respective orthogonal
planes (Fig. 53).

Emission in the x- and y-directions is linearly polarized in
the x$\underline{E}$ and y$\underline{E}$ planes, respectively, with components of intensity
polarized parallel to the z-, y-, and x-coordinates of given by

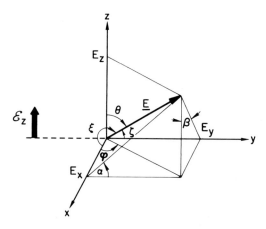

FIG. 53. Emission transition moment vector in excitation-based
coordinate system.

$$I_z = \sin^2 \xi \, \sin^2 \alpha = \sin^2 \zeta \, \cos^2 \beta$$

$$I_y = \sin^2 \xi \, \cos^2 \alpha \qquad\qquad\qquad\qquad (A2)$$

$$I_x = \sin^2 \zeta \, \sin^2 \beta$$

From Fig. 53 it is readily seen that

$$\sin \xi \, \sin \alpha = \sin \zeta \, \cos \beta = \cos \theta$$

$$\sin \xi \, \cos \alpha = \cos \zeta = \sin \theta \, \sin \varphi \qquad\qquad (A3)$$

$$\cos \xi = \sin \zeta \, \sin \beta = \sin \theta \, \cos \varphi$$

By using Eqs. (A2) and (A3) in (A1), r_p may be written in alternative ways:

$$r_p(\xi,\alpha) = \sin^2 \xi (2 \sin^2 \alpha - 1)$$

$$r_p(\zeta,\beta) = \sin^2 \zeta (1 + \cos^2 \beta) - 1 \qquad\qquad (A4)$$

$$r_p(\theta,\varphi) = \cos^2 \theta (1 + \sin^2 \varphi) - \sin^2 \varphi$$

For the applications of polarized emission spectroscopy discussed in the present chapter, the relevant EA is that corresponding to a random, i.e., isotropic, distribution of absorption transition dipoles. Since the absorption probability depends on the orientation of the absorption transition moment vector with respect to the polarization vector of excitation, a photoselection of the absorption dipoles and therefore, in general, of the associated emission dipoles occurs, the observed EA being an average value over the polarized absorption probability. Consider an emission vector $\underline{E}$ oriented at some angle ω to the associated absorption vector $\underline{A}$ and characterized by an azimuthal angle δ with respect to the latter which itself is defined by the Euler and azimuthal angles θ' and φ', respectively (also by ξ', ζ', and α', β') (Fig. 54). These angles are related by

$$\cos \theta = \cos \theta' \, \cos \omega + \sin \theta' \, \sin \omega \, \cos \delta \qquad\qquad (A5a)$$

By squaring Eq. (A5a) and by averaging $\cos^2 \theta$ over the azimuthal

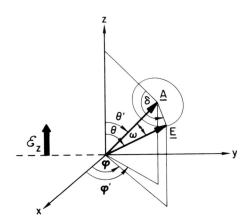

FIG. 54. Absorption and emission transition moment vectors in excitation-based coordinate system.

angle δ (since in isotropic solutions all values of δ are equally likely for any given θ' and ω, $<\cos \delta> = 0$ and $<\cos^2 \delta> = 1/2$,

$$\cos^2 \theta = \cos^2 \theta' \cos^2 \omega + \frac{1}{2} \sin^2 \theta' \sin^2 \omega \tag{A5b}$$

Substitution of this result into Eq. (A4) similarly averaged over the azimuthal angle φ leads, after some rearrangement, to

$$r_p(\theta',\omega) = \left(\frac{3}{2} \cos^2 \theta - \frac{1}{2}\right) = \left(\frac{3}{2} \cos^2 \theta' - \frac{1}{2}\right) \left(\frac{3}{2} \cos^2 \omega - \frac{1}{2}\right)$$

$$\tag{A6}$$

For an isotropic distribution of absorption transition moment vectors, the excitation probability for z-polarized excitation is independent of the azimuthal angle φ so that z- is a symmetry axis for the distribution of excited transition moment vectors and all azimuths, φ, are equally likely in this distribution. The first . term in θ' can be identified with the EA of the absorption vector, i.e., that which would characterize fluorescence with no orientation change after absorption, while the second is a depolarization term arising from such a reorientation.

2. Reorientation and Depolarization Factors

It is readily shown by induction that, quite generally, if any emission transition moment vector $\underline{E}_i$ is azimuthally randomly distributed about the vector direction $\underline{E}_{i-1}$, the EA may be written as a product of depolarizing factors d_i:

$$r_p = \prod_i \left(\frac{3}{2} \cos^2 \theta_i - \frac{1}{2} \right) = \prod_i d_i \qquad (A7)$$

where θ_i is the angle between $\underline{E}_i$ and $\underline{E}_{i-1}$.

The first such depolarizing factor, involving $\theta_1 = \theta'$ in Eq. (A6), is that produced by the absorption process itself. For an isotropic solution of absorbers, the average of $\cos^2 \theta'$ over the absorption probability, which is itself proportional to $\cos^2 \theta'$, is appropriate and is given by

$$<\cos^2 \theta'> = \frac{\int_0^{\pi} \cos^4 \theta' \sin \theta' \, d\theta'}{\int_0^{\pi} \cos^2 \theta' \sin \theta' \, d\theta'} = \frac{3}{5} \qquad (A8)$$

Therefore,

$$r_{p,absn} = \frac{3}{2} <\cos^2 \theta'> - \frac{1}{2} = 0.4 \qquad (A9)$$

When $\omega = 0$, the absorption and emission dipole transition moment vectors are the same (or parallel), and if they do not change their orientation between absorption and emission, the EA has the same (limiting) value, 0.4. When ω has some arbitrary value, the EA is diminished, and in the absence of motion or other depolarizing factors, is referred to as the fundamental EA, r_f:

$$r_f(\omega) = 0.4 \left(\frac{3}{2} \cos^2 \omega - \frac{1}{2} \right) \qquad (A10)$$

where

$$-0.2 \leq r_f(\omega) \leq 0.4 \tag{A11}$$

linearly polarized excitation being understood.

It can also be seen from Eqs. (A2) and (A3) that

$$I_y = \sin^2 \theta \sin^2 \varphi$$

and $\hspace{10cm}$ (A12)

$$I_x = \sin^2 \theta \cos^2 \varphi$$

By averaging over the azimuthal angle $\omega (<\cos^2 \varphi> = <\sin^2 \varphi> = \frac{1}{2})$, it is clear that I_x and I_y are equal and that they are also equal to the component of emission intensity polarized perpendicular to z and observed anywhere in the xy plane. The EA defined in Eq. (A1) becomes

$$r_p = \frac{I_z - I_y}{I_z + 2I_y} \tag{A13a}$$

When linearly polarized excitation is used, it is conventional to observe the emission in the plane orthogonal to the excitation polarization direction and to refer the polarized emission intensities to this direction. The emission intensity polarized parallel to it, I_z, then becomes $I_{\parallel}$, while the component polarized perpendicular to it, $I_{\perp}$, replaces I_y, for observation anywhere in the plane perpendicular to the excitation polarization vector. The EA may then be rewritten as

$$r_p = \frac{I_{\parallel} - I_{\perp}}{I} = \frac{I_{\parallel} - I_{\perp}}{I_{\parallel} + 2I_{\perp}} \tag{A13b}$$

3. Emission Anisotropy in the Laboratory Coordinate System

In the laboratory it is usually convenient to have the exciting light incident in the horizontal plane and also to observe the emission in this plane and at right angles to the excitation direction. Under these conditions it is useful to assign an excitation anisotropy r_{ex} to the exciting light, defined in the vertical (V) and horizontal (H) laboratory coordinates by

$$r_{ex} = \frac{I_{Vex} - I_{Hex}}{I_{ex}}$$

(A14)

where I_{Vex} is the component of excitation polarized in the laboratory V-direction, I_{Hex} the component in the horizontal plane at right angles to the observation direction, and I_{ex} is the total intensity. The excitation anisotropy is then fully defined by the cosine square of the angle θ_{ex} between the given polarization vector and the V-axis, i.e., since $I_{Vex} = \cos^2 \theta_{ex}$ and $I_{Hex} = 0$,

$$r_{ex} = \cos^2 \theta_{ex}$$

(A15)

By resolving the excitation into its vertical and horizontal components in the plane orthogonal to the propagation direction, the relative polarized emission intensities may be expressed as

$$I_V = \cos^2 \theta_{ex} I_{\parallel} + \sin^2 \theta_{ex} I_{\perp}$$

(A16a)

and

$$I_H = \cos^2 \theta_{ex} I_{\perp} + \sin^2 \theta_{ex} I_{\perp} = I_{\perp}$$

(A16b)

Substituting these two relationships into Eq. (12), the EA is seen to be

$$r = \cos^2 \theta_{ex} \left(\frac{I_{\parallel} - I_{\perp}}{I} \right)$$

(A17)

which, with Eqs. (A13b) and (A15) leads to

$$r = r_{ex} r_p$$

(A18)

This equation permits expression of the EA of an isotropic absorbing sample irradiated by light of any polarization simply in terms of the anisotropy of the exciting light and the EA obtained with linearly polarized excitation, provided the total intensity I is measured appropriately as the sum of three mutually orthogonal polarized intensities.

By writing down the third orthogonally polarized component of emission, $I_{H'} = \cos^2 \theta_{ex} I_\perp + \sin^2 \theta_{ex} I_\parallel$, it is readily seen to be equal to I_H when $\theta_{ex} = 0$ (excitation fully V-polarized) and to I_V when $\cos^2 \theta_{ex} = 1/2$. The latter condition is realized with so-called _natural_ (unpolarized) excitation and with circularly polarized excitation (and, of course, excitation linearly polarized at 45 deg to the vertical laboratory axis). Light emerging from a monochromator, or even light selected by filters directly from a ribbon filament lamp that does not have a plane exit window, is generally quite highly polarized, necessitating the interposition of a correctly oriented $\frac{1}{4}$ wave scrambler plate in the excitation beam for natural excitation spectroscopy. Thus, omitting the subscript p, the EA's for polarized and _natural_ excitation are given by

$$r = \frac{I_V - I_H}{I_V + 2I_H} \quad \text{(vertically polarized excitation)} \tag{A19a}$$

$$r_n = \frac{I_V - I_H}{2I_V + I_H} \quad \text{(natural excitation)} \tag{A19b}$$

and

$$r_n = r/2 \tag{A19c}$$

The condition $r_{ex} = 0$, corresponding to horizontally (H'-) polarized excitation, is useful as a reference in measurement of I_V and I_H which are seen from the above considerations to be rendered equal under this condition [45].

4. Degree of Polarization

Classically, luminescence polarization has been defined in terms of the so-called degree of polarization, p, e.g. [22]:

$$p = \frac{I_V - I_H}{I_V + I_H} \tag{A20}$$

The relationship between p and the EA is obtained by comparing Eqs. (A19) and (A20),

$$r = \frac{2p}{3 - p} \quad \text{and} \quad r_n = \frac{2p_n}{3 + p_n} \tag{A21a}$$

while $p_n = p/(2 - p)$ corresponds to $r_n = r/2$.

By using the derived reciprocal relationships

$$r^{-1} = \frac{3}{2}\left(\frac{1}{p} - \frac{1}{3}\right) \quad \text{and} \quad r_n^{-1} = \frac{3}{2}\left(\frac{1}{p} + \frac{1}{3}\right)$$

it is possible to state Perrin's law of isotropic depolarization [68] and Weber's law of addition of polarizations [46] in greatly simplified forms.

5. Perrin's Law of Isotropic Depolarization

This law relating to the isotropic Brownian rotational movements of luminophores in nonrigid media is easily derived from Eqs. (A6), (A7), and (A10). If θ_j are M (small) azimuthally averaged angular displacements occurring independently of each other, then [38,69]

$$\frac{r}{r_f} = \prod_{j=1}^{M} \left(\frac{3}{2} \cos^2 \theta_j - \frac{1}{2}\right) = \frac{3}{2} \cos^2 \theta - \frac{1}{2} \tag{A22}$$

where θ is the angle between the initial direction of the emission moment vector and its direction at the moment of emission. By averaging over all θ,

$$\frac{\langle r \rangle}{r_f} = \frac{3}{2} \langle \cos^2 \theta \rangle - \frac{1}{2} \tag{A23a}$$

which may be recast in terms of degree of polarization by using Eq. (A21b) to give

$$\left(\frac{1}{p} \mp \frac{1}{3}\right) = \left(\frac{1}{p_o} \mp \frac{1}{3}\right)\left(\frac{2}{3<\cos^2 \theta> - 1}\right) \tag{A23b}$$

which is Perrin's law of isotropic depolarization, p_o being the value of p extrapolated to zero temperature and infinite viscosity (which should be equivalent to the fundamental polarization, see discussion in Sec. IV). The value of $<\cos^2 \theta>$ can be related to the isotropic rotational relaxation times of the luminophore itself or a larger molecular substrate to which it may be attached either rigidly or with some degree of freedom.

6. Weber's Law of Addition of Polarizations

By rearranging Eq. (12) for the ith species of N emitters contributing an intensity I_i to the total intensity I,

$$r_i I_i = I_{Vi} - I_{Hi} \tag{A24}$$

By summing over the N species and dividing by the total intensity $I = \Sigma_{i=1}^{N} I_i$, the average EA is given by

$$<r> = \sum_{i=1}^{N} r_i I_i / \sum_{i=1}^{N} I_i$$

$$= \sum_{i=1}^{N} f_i r_i \tag{A25a}$$

where f_i is the fraction of the total intensity contributed by the ith species, i.e., $f_i = I_i/I$. By substituting the relationships of Eq. (A21b) into the last and rearranging

$$<\frac{1}{p} \mp \frac{1}{3}> = \left(\sum_{i=1}^{N} \frac{f_i}{\frac{1}{p_i} \mp \frac{1}{3}}\right)^{-1} \tag{A25b}$$

which is explicitly Weber's addition law for polarizations, the
upper and lower signs referring to polarized and natural excitation,
respectively.

Another polarization addition law, e.g. [22,46],

$$<p> = \sum_{i=1}^{N} p_i I'_i / \sum_{i=1}^{N} I'_i \tag{A26}$$

where I' represents $(I_V + I_H)$, i.e., the sum of the V- and H-polar-
ized components seen in the direction of viewing, is also applicable
but relatively unwieldy since, except for $p = 0$, I'_i is not directly
related to the mole fraction of species i.

Weber has given a short review of polarized emission spectroscopy [70], and, more recently, a detailed review of its principles
and practice has become available [71].

A discussion of time-dependent depolarization phenomena arising, e.g., from rotation or energy transfer with correlation times
of the same order as the emission lifetime, is not given here.
However, the former is discussed briefly in connection with the
analysis of experimental data on transfer depolarization (Sec. IV).
An extensive account of the theory and practice of the decay of
emission anisotropy is given elsewhere in this volume [29].

APPENDIX B: THE ORIENTATION FACTOR AND
TRANSFER DEPOLARIZATION — ILLUSTRATIVE SELECTED MODELS

In Sec. III, general expressions were given for the average
values of the orientation factor κ^2 and for those of $\cos^2 \theta_T$ which
determine the transfer depolarization d_T under dynamic and static
conditions [Eqs. (24)-(27)]. For illustrative purposes, simple
geometrical models for which the averages of these parameters may
be obtained as functions of either one or two angles have been selected. As already indicated, the frequency factor n_i for a given

configuration i has been taken as a purely geometrical weighting
element over the region corresponding to the orientational freedom
of either or both donor and acceptor and the averaging performed by
integrating over this region.

Model 1 (Fig. 5). One transition moment is fixed at an angle θ
to the separation vector R; the other has a random orientation. Averaging is thus carried out for the second moment which makes an
angle θ' with R over all θ' and all azimuthal angles φ with weighting
element $n_1 = \sin\theta'\, d\theta'\, d\varphi$. Using this element and substituting Eq.
(5) into (27), Eq. (22) into (25), and Eqs. (5) and (22) into (26)
gives

$$\langle \kappa^2 \rangle = \frac{\int_{\theta'=0}^{\pi}\int_{\varphi=0}^{2\pi}(\sin\theta\,\sin\theta'\,\cos\varphi - 2\cos\theta\,\cos\theta')^2\,\sin\theta'\,d\theta'\,d\varphi}{\int_{\theta'=0}^{\pi}\int_{\varphi=0}^{2\pi}\sin\theta'\,d\theta'\,d\varphi}$$

$$\langle \cos^2\theta_T \rangle = \frac{\int_{\theta'=0}^{\pi}\int_{\varphi=0}^{2\pi}(\sin\theta\,\sin\theta'\,\cos\varphi + \cos\theta\,\cos\theta')^2\,\sin\theta'\,d\theta'\,d\varphi}{\int_{\theta'=0}^{\pi}\int_{\varphi=0}^{2\pi}\sin\theta'\,d\theta'\,d\varphi}$$

and

$$\lim_{\langle T \rangle \to 0} \langle \cos^2\theta_T \rangle_s = \frac{\langle \kappa^2 \cos^2\theta_T \rangle}{\langle \kappa^2 \rangle}$$

where

$$\langle \kappa^2 \cos^2\theta_T \rangle$$

$$= \frac{\int_{\theta'=0}^{\pi}\int_{\varphi=0}^{2\pi}(\sin\theta\sin\theta'\cos\varphi - 2\cos\theta\cos\theta')^2\,(\sin\theta\sin\theta'\cos\varphi + \cos\theta\cos\theta')^2\sin\theta'\,d\theta'\,d\varphi}{\int_{\theta'=0}^{\pi}\int_{\varphi=0}^{2\pi}\sin\theta'\,d\theta'\,d\varphi}$$

By multiplying out the factors and performing the integrations,

$$<\kappa^2> = \frac{1}{3} (1 + 3 \cos^2 \theta)$$

$$<\cos^2 \theta_T>_d = \frac{1}{3} \text{ [leading to } <d_T>_d = 0, \text{ independent of } \theta]$$

$$<\kappa^2 \cos^2 \theta_T> = \frac{1}{5} (1 - 3 \cos^2 \theta + 6 \cos^4 \theta)$$

These expressions were used to calculate numerical values for $<\kappa^2>$, $<d_T>_d$ and $\lim_{<T> \to 0} <d_T>_s$ which are displayed as functions of θ in Fig. 5.

Model 2 (Fig. 6). The two transition moments are constrained to lie in planes which intersect along R and subtend an angle φ. The weighting element in this case is $(d\theta \, d\theta')$, and the relevant averages are given by

$$<\kappa^2> = \frac{1}{4} (4 + \cos^2 \varphi)$$

$$<\cos^2 \theta_T>_d = \frac{1}{4} (1 + \cos^2 \varphi)$$

$$<\kappa^2 \cos^2 \theta_T> = \frac{3}{64} (12 - \cos^2 \varphi + 3 \cos^4 \varphi)$$

Model 3 (Fig. 7). One transition moment makes an angle θ with the separation vector R, the other being constrained to lie in a plane which subtends an angle φ with the plane containing the first moment and R. The average value of κ^2 and those determining $<d_T>_d$ and $\lim_{<T> \to 0} <d_T>_s$ are

$$<\kappa^2> = \frac{1}{2} (1 + 3 \cos^2 \theta - \sin^2 \theta \sin^2 \varphi),$$

$$<\cos^2 \theta_T>_d = \frac{1}{2} (1 - \sin^2 \theta \sin^2 \varphi),$$

$$<\kappa^2 \cdot \cos^2 \theta_T> = \frac{3}{8} (\sin^4 \theta \cos^4 \varphi - \sin^2 \theta \cos^2 \theta \cos^2 \varphi + 4 \cos^4 \theta)$$

The average values of the orientation factor and depolarization factors as functions of θ and φ are displayed in Fig. 7, both as three-dimensional plots in true perspective and as contour plots.

Model 4 (Figs. 8-16). One transition moment is fixed at an angle θ with respect to the separation vector $\underline{R}$; the other is constrained to the surface or volume of a cone. The axis of the cone lies either (1) along $\underline{R}$, (2) perpendicular to $\underline{R}$ and in the same plane as the fixed transition moment, or (3) perpendicular to this plane. The relevant weighting elements for the cone surface distribution (c) are (see Fig. 55): (i) $d\varphi$, (ii) and (iii) $d\gamma$, and for the volume distribution (C): (i) $\sin \theta' \, d\theta' \, d\varphi$, (ii) and (iii) $\sin \psi \, d\psi \, d\gamma$ where the cone half-angle ψ and associated azimuth γ in (ii) are related to the angles θ' and φ appearing in the expressions for κ^2 and d_T by

$$\cos \theta' = \sin \psi \sin \gamma$$

$$\sin \theta' \sin \emptyset = \sin \psi \cos \gamma$$

$$\sin \theta' \cos \emptyset = \cos \psi$$

where $\emptyset = \varphi$ (Fig. 55) and in (iii) by

$$\cos \theta' = \sin \psi \sin \gamma$$

$$\sin \theta' \cos \emptyset = \sin \psi \cos \gamma$$

$$\sin \theta' \sin \emptyset = \cos \psi$$

where $\emptyset = \pi/2 - \varphi$.

By integrating over the azimuthal angles φ, $\emptyset$, or γ, the averages in the surface distribution (c) cases are given by

(i,c) $<\kappa^2> = \frac{1}{2} s^2\theta_1 s^2\theta_2 + 4c^2\theta_1 c^2\theta_2$

$<\cos^2 \theta_T>_d = \frac{1}{2} s^2\theta_1 s^2\theta_2 + c^2\theta_1 c^2\theta_2$

$<\kappa^2 \cos^2 \theta_T> = \frac{3}{8} s^4\theta_1 s^4\theta_2 - \frac{3}{2} c^2\theta_1 s^2\theta_1 c^2\theta_2 s^2\theta_2 + 4c^4\theta_1 c^4\theta_2$

with θ_1,θ_2 replacing θ,θ' and using c, s for cos, sin. Numerical values of $<\kappa^2>$, $<d_T>_d$, and $\lim_{<T>\to 0} <d_T>_s$ derived from these equations are displayed as functions of θ_1 and θ_2 in three-dimensional and contour plots in Fig. 8.

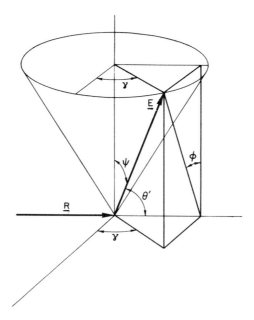

FIG. 55. Cone geometry for quadrature Models 4 and 5.

(ii,c) $\quad <\kappa^2> = s^2\theta c^2\psi + 2c^2\theta^2 s^2\psi$

$\quad\quad <\cos^2 \theta_T>_d = s^2\theta c^2\psi + \frac{1}{2} c^2\theta s^2\psi$

$\quad\quad <\kappa^2 \cos^2 \theta_T> = s^4\theta c^4\psi - \frac{3}{2} c^2\theta s^2\theta c^2\psi s^2\psi + \frac{3}{2} c^4\theta s^4\psi$

(See Fig. 11.)

(iii,c) $\quad <\kappa^2> = \frac{1}{2} s^2\psi(1 + 3c^2\theta)$

$\quad\quad <\cos^2 \theta_T> = \frac{1}{2} s^2\psi$

$\quad\quad <\kappa^2 \cos^2 \theta_T> = \frac{3}{8} s^4\psi(s^4\theta - c^2\theta s^2\theta + 4c^4\theta)$

(See Fig. 14.)

The averages in the volume distribution cases (C) may be obtained directly from the surface distribution averages by integrating the latter over the relevant volume using the weighting elements sin θ dθ and sin ψ dψ, i.e., by replacing the functions of θ and ψ by their averages over this volume. The half-cone angles Θ, Ψ now represent the upper limits for integration, and

$$<c^2\psi>_\Psi = \frac{1}{3}(1 + c\Psi + c^2\Psi)$$

$$<s^2\psi>_\Psi = 1 - <c^2\psi>_\Psi$$

$$<c^4\psi>_\Psi = \frac{1}{5}(1 + c\Psi + c^2\Psi + c^3\Psi + c^4\Psi)$$

$$<s^4\psi>_\Psi = \frac{1}{15}(8-7[c\Psi + c^2\Psi] + 3[c^3\Psi + c^4\Psi])$$

$$<c^2\psi s^2\psi>_\Psi = \frac{1}{15}(2[1 + c\Psi + c^2\Psi] - 3[c^3\Psi + c^4\Psi])$$

while the average values $<f(\theta)>_\Theta$ are defined analogously.

Cases (i,C)-(iii,C) are presented in Figs. 9, 12, and 15 for $0 \le \Theta, \Psi \le \pi/2$. In cases (c) there is a plane of symmetry for the values of $<K^2>$, $<d_T>_d$, and $\lim_{<T> \to 0} <d_T>_s$ which is defined by $\theta, \psi = \pi/2$. In cases (C), on the other hand, the situation is different. As the cone becomes larger than a hemisphere, the average values reflect the absence of distribution over the volume of a cone of half-angle $\Theta'\psi'$ less than $\pi/2$. These cases (C') are described by replacing $<f(\theta)>_\Theta$, $<f(\psi)>_\Psi$ in the above expressions by $<f(\theta)>_{\Theta'}$, $<f(\psi)>_{\Psi'}$, where $\Theta' = (\pi - \Theta)$ and $\Psi' = (\pi - \Psi)$ in the interval $[\pi/2 \le \Theta, \Psi \le \pi]$ in which case $\cos \Theta$, $\cos \Psi$ in the expressions above are replaced by $(-\cos \Theta')$, $(-\cos \Psi')$. The results for these cases, (i,C')-(iii,C') are given in Figs. 10, 13, and 16 as functions of Θ', Ψ'.

Model 5 (Figs. 8-10 and 17-40). The two transition moments are restricted to the surface (c) or volume (C or C') of cones, and the cone axes may be parallel or perpendicular to each other and lie either along R or perpendicular to it. The four possible relative orientations of the cone axes are then: (i) parallel in line (both along R), (ii) parallel in plane (both perpendicular to R), (iii) perpendicular in plane (one along R), (iv) perpendicular in perpendicular planes (both perpendicular to R).

The expressions of Eqs. (5) and (22) for κ^2 and $\cos^2 \theta_T$ are rewritten where appropriate in terms of the cone half-angles and azimuths as above with $\varphi = (\phi_2 - \phi_1)$, taking the plane containing $\underline{R}$ and the axis of cone 1 as the reference plane for both ϕ_2 and ϕ_1. Clearly case (i) is equivalent to case (i) of Model 3, and the rest of these cases are described by

(ii) $\quad <\kappa^2> = c^2\psi_1 c^2\psi_2 + \frac{5}{4} s^2\psi_1 s^2\psi_2$

The equivalent of this expression has been given previously for donors and acceptors in a monolayer at a solvent-air interface where the transition moments are inclined at an angle to the surface, but all azimuths about the normal to the surface are possible [72].

$$<\cos^2 \theta_T> = c^2\psi_1 c^2\psi_2 + \frac{1}{2} s^2\psi_1 s^2\psi_2$$

$$<\kappa^2 \cos^2 \theta_T> = c^4\psi_1 c^4\psi_2 + \frac{3}{4} c^2\psi_1 s^2\psi_1 c^2\psi_2 s^2\psi_2 + \frac{21}{32} s^4\psi_1 s^4\psi_2$$

(iii) $\quad <\kappa^2> = \frac{1}{4} s^2\theta(1 + c^2\psi) + 2c^2\theta s^2\psi$

$$<\cos^2\theta_T>_d = \frac{1}{4} s^2\theta(1 + c^2\psi) + \frac{1}{2} c^2\theta s^2\psi$$

$$<\kappa^2 \cos^2\theta_T> = \frac{3}{8} (s^4\theta[c^4\psi + c^2\psi s^2\psi + \frac{3}{8} s^4\psi]$$

$$- c^2\theta s^2\theta[\frac{1}{2} s^4\psi + 2c^2\psi s^2\psi] + 4c^4\theta s^4\psi)$$

(iv) $\quad <\kappa^2> = \frac{1}{2} (\sin^2\psi_1 + \sin^2\psi_2)$

$$<\cos^2\theta_T>_d = \frac{1}{2} (\sin^2\psi_1 + \sin^2\psi_2 - \frac{3}{2} \sin^2\psi_1 \sin^2\psi_2)$$

$<\cos^2\theta_T>_d$ in (iii) and (iv) rearrange to the same expression in two variables, say α and β:

$$<\cos^2\theta_T>_d = \frac{1}{4} [1 + c^2\alpha + c^2\beta - 3c^2\alpha c^2\beta]$$

but are presented here in the form directly derived.

$$
\begin{aligned}
<\kappa^2 \cos^2 \theta_T> = \frac{3}{2} \Bigg[& \left(c^4\psi_1 s^4\psi_2 + s^4\psi_1 c^4\psi_2 \right) \\
& - \frac{1}{8} \left(c^2\psi_1 s^2\psi_1 s^4\psi_2 + s^4\psi_1 c^2\psi_2 s^2\psi_2 \right) \\
& + c^2\psi_1 s^2\psi_1 c^2\psi_2 s^2\psi_2 + \frac{1}{8} s^4\psi_1 s^4\psi_2 \Bigg]
\end{aligned}
$$

Numerical results for cases (i) with one of the transition moments having a cone surface distribution only, the other being surface [case (i,cc)] or volume [cases (i,cC), (i,cC′)] distributed, are already presented graphically in Figs. 8-10. The other combinations of surface and volume distribution for this case, i.e. (i,CC), (i,CC′), and (i,C′C′), are given in Figs. 17-19. The rest, i.e., cases (ii)-(iv), are given in Figs. 20-40, and a summary of the possible combinations for cases (i)-(iv) is presented in Table 2.

Model 6 (Figs. 41-45). The two transition moment vectors are each constrained to lie in planes (P_1, P_2) whose normals make angles ($\pi/2 - \Theta_1$), ($\pi/2 - \Theta_2$) with R and lie in planes which are disposed at an angle Φ to each other. If γ is the angle between a transition moment E lying in the plane of constraint P and the intersection E' of P with the plane P_n containing its normal N and the separation vector R, then the angle θ determining the orientation of E with respect to R and the angle $\emptyset$ between P_n and the plane P_t containing both E and R are related to Θ (i.e., Θ_1 or Θ_2 defined above) and γ by

$$\cos \theta = \cos \Theta \cos \gamma$$

$$\sin \theta \cos \emptyset = \sin \Theta \cos \gamma$$

$$\sin \theta \sin \emptyset = \sin \gamma$$

(See Fig. 56.)

The expressions for κ^2 and $\cos^2 \theta_T$ [Eqs. (5) and (22)] may now be rewritten in terms of $\Theta_{1,2}$, $\gamma_{1,2}$, and Φ taking P_{n_1} as the reference plane for Φ so that $\varphi = (\Phi + \emptyset_2 - \emptyset_1)$ in Eqs. (5) and (22). Figures 41-43 display $<\kappa^2>$, $<d_T>_d$, and $\lim_{<T> \to 0} <d_T>_s$ as functions of Θ_1 and Θ_2 for cases (i)-(iii) in which Φ takes on values of 0,

TABLE 2

Combinations of Surface and
Volume Distributions in Model 5

The numerals 1 and 2 represent the two transition moments while
θ,ψ indicate distributions over conical surfaces, Θ,Ψ distributions
over conical volumes in the limits $0 \leq \Theta,\Psi \leq \pi/2$, and $\Theta'\Psi'$ those
over conical volumes in the limits $\pi/2 \leq \Theta,\Psi \leq \pi$. Combinations are
written as cc for surface-surface, cC for surface-volume (up to hemisphere), cC' for surface-volume (greater than hemisphere), etc., and
the number is that of the figure displaying $<\kappa^2>$, $<d_T>_d$, and
$\lim_{<T>\to 0} <d_T>_s$ as functions of the cone angles.

Case (i)

1 \ 2	θ	Θ	Θ'
θ	cc 8	cC 9	cC' 10
Θ		CC 17	CC' 18
Θ'			C'C' 19

Case (ii)

1 \ 2	ψ	Ψ	Ψ'
ψ	cc 20	cC 21	cC' 22
Ψ		CC 23	CC' 24
Ψ'			C'C' 25

TABLE 2 (Continued)

Case (iii)

	θ	⊖	⊖′
ψ	cc 26	cC 27	cC′ 29
Ψ	Cc 28	CC 31	CC′ 32
Ψ′	C′c 30	C′C 33	C′C′ 34

Case (iv)

1 \ 2	ψ	Ψ	Ψ′
ψ	cc 35	cC 36	cC′ 37
Ψ		CC 38	CC′ 39
Ψ′			C′C′ 40

In cases (i), (ii), and (iv), Cc, C′c, and C′C are equivalent to cC, cC′, and CC′, respectively.

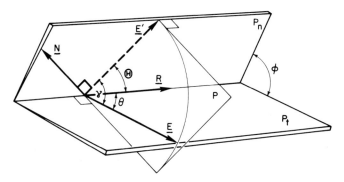

FIG. 56. Geometry for intercalation Models 6.

$\pi/2$, and π, respectively. The results obtained when $\Theta_1 = \Theta_2 = \Theta$ and Φ is variable, case (iv) with $0 \leq \Phi \leq \pi/2$ and case (v) with $\pi/2 \leq \Phi \leq \pi$, are presented in Figs. 44 and 45, respectively.

The equations governing these cases after averaging over $\gamma_{1,2}$ by integration over the planes P_1 and P_2 with weighting factors $d\gamma_{1,2}$ are

(i) $\Phi = 0$ (Fig. 41):

$$< \kappa^2 > = \frac{1}{4} [1 + (s\Theta_1 s\Theta_2 - 2c\Theta_1 c\Theta_2)^2]$$

$$< \cos^2 \theta_T >_d = \frac{1}{4} [1 + (s\Theta_1 s\Theta_2 + c\Theta_1 c\Theta_2)^2]$$

$$< \kappa^2 \cos^2 \theta_T > = \frac{3}{64} [3(s^4\Theta_1 s^4\Theta_2 - 2s^3\Theta_1 c\Theta_1 s^3\Theta_2 c\Theta_2$$
$$- 3s^2\Theta_1 c^2\Theta_1 s^2\Theta_2 c^2\Theta_2 + 4s\Theta_1 c^3\Theta_1 s\Theta_2 c^3\Theta_2$$
$$+ 4c^4\Theta_1 c^4\Theta_2) + 2(s^2\Theta_1 s^2\Theta_2 - s\Theta_1 c\Theta_1 s\Theta_2 c\Theta_2)$$
$$- c^2\Theta_1 c^2\Theta_2 + 3]$$

(ii) $\Phi = \pi/2$ (Fig. 42):

$$< \kappa^2 > = \frac{1}{4} [s^2\Theta_1 + s^2\Theta_2 + 4c^2\Theta_1 c^2\Theta_2]$$

$$< \cos^2 \theta_T >_d = \frac{1}{4} [1 + s^2\Theta_1 s^2\Theta_2]$$

$$< \kappa^2 \cos^2 \theta_T > = \frac{3}{64} [3s^4\Theta_1 + 2s^2\Theta_1 s^2\Theta_2 + 3s^4\Theta_2 - 3(c^2\Theta_1 c^2\Theta_2 s^2\Theta_2$$
$$+ c^2\Theta_1 s^2\Theta_1 c^2\Theta_2) + 12c^4\Theta_1 c^4\Theta_2]$$

(iii) $\Phi = \pi$ (Fig. 43): The equations are as in (i) with the sign of terms containing odd powers of $\cos \Theta_{1,2}$ reversed.

(iv) $0 \leq \Phi \leq \pi/2$ (Fig. 44):

$$< \kappa^2 > = \frac{1}{4} [c^2\Phi + 2s^2\Phi s^2\Theta + (s^2\Theta c\Phi - 2c^2\Theta)^2]$$

$$< \cos^2 \theta_T >_d = \frac{1}{4} [c^2\Phi + 2s^2\Phi s^2\Theta + (s^2\Theta c\Phi + c^2\Theta)^2]$$

$$\langle \kappa^2 \cos^2 \theta_T \rangle = \frac{3}{64} [c^4\Phi(3s^8\Theta + 2s^4\Theta + 3)$$

$$+ 4c^2\Phi s^2\Phi s^2\Theta(3s^4\Theta - 2s^2\Theta + 3)$$

$$+ 8s^4\Phi s^4\Theta - 2c^3\Phi c^2\Theta s^2\Theta(3s^4\Theta + 1)$$

$$- 4c\Phi s^2\Phi c^2\Theta s^2\Theta(3s^2\Theta - 1) - c^2\Phi c^4\Theta(9s^4\Theta + 1)$$

$$- 6s^2\Phi c^4\Theta s^2\Theta + 12c\Phi c^6\Theta s^2\Theta + 12c^8\Theta]$$

(v) $\pi/2 \leq \Phi \leq \pi$ (Fig. 45): Equations as in (iv) with Φ replaced by $(\pi - \Phi)$, i.e., sign of terms in odd powers of $\cos \Phi$ reversed.

Model 7 (Figs. 46-51). Local minima and maxima in κ^2 are observed for particular sets of relative orientation of D and A (see Sec. II). The same is true in general if either or both D and A have some orientational freedom. Two examples are presented for illustration here. In both of them, one of the transition moments $\underline{E}_1$ is rigidly fixed at an angle θ with respect to the separation vector $\underline{R}$, the other having reorientational freedom over the surface or volume of cones of half-angles ψ and Ψ, respectively.

In the first example the axis $\underline{C}$ of the cone is disposed at an azimuth Φ of zero about the $\underline{E}_1,\underline{R}$ plane (i.e., in the same plane) and makes an angle ξ with the separation vector (see, e.g., Fig. 46) related to θ by

$$\sin \theta \sin \xi - 2 \cos \theta \cos \xi = 0$$

corresponding to a zero value of κ for a rigid transition moment along $\underline{C}$ [$\psi = 0$, $\varphi = \Phi = 0$, see Eq. (5)]. Defining an arbitrary vector $\underline{E}_2$ of the cone distribution by its polar and azimuthal angles θ_c and ϕ_c with respect to $\underline{R}$ and the $\underline{E}_1,\underline{R}$ plane, and by its polar angle ψ about $\underline{C}$ and associated azimuth γ in the plane orthogonal to $\underline{C}$ as in previous cases, then (see Fig. 57)

$$\sin \theta_c \cos \phi_c = \sin \xi \cos \psi - \cos \xi \sin \psi \cos \gamma$$

$$\cos \theta_c = \cos \xi \cos \psi + \sin \xi \sin \psi \cos \gamma$$

$$\sin \theta_c \sin \phi_c = \sin \psi \sin \gamma$$

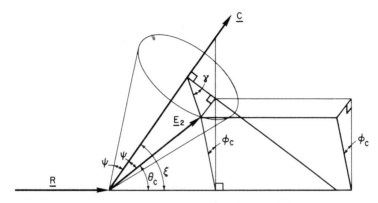

FIG. 57. Cone geometry for minima/maxima Models 7.

By substituting these expressions into those for κ^2, $\cos^2 \theta_T$, and $\kappa^2 \cdot \cos^2 \theta_T$ with $\varphi = \phi_c$ and performing the relevant integrations, the average values for the surface distribution case (i,c) are given by

$$<\kappa^2> = \frac{1}{2} \sin^2 \psi (1 + 3 \cos^2 \theta)$$

$$<\cos^2 \theta_T>_d = 9 \cos^2 \psi \left(\frac{\cos^2 \theta \, \sin^2 \theta}{1 + 3 \cos^2 \theta}\right)$$

$$+ \frac{1}{2} \sin^2 \psi \left(\frac{[1 - 3 \cos^2 \theta]^2}{1 + 3 \cos^2 \theta}\right)$$

$$\lim_{<T> \to 0} <\cos^2 \theta_T>_s = 9 \cos^2 \psi \left(\frac{\cos^2 \theta \, \sin^2 \theta}{1 + 3 \cos^2 \theta}\right)$$

$$+ \frac{3}{4} \sin^2 \psi \left(\frac{[1 - 3 \cos^2 \theta]^2}{1 + 3 \cos^2 \theta}\right)$$

This result is displayed in Fig. 46. The volume averaging equivalents (i,C) and (i,C') are obtained as before by substituting average values of functions of ψ into the surface average equations, and these cases are depicted in Figs. 47 and 48 for cone half-angles less than or greater than $\pi/2$. It can be shown that $<\kappa^2>$ has a local minimum value when $\cos^2 \psi$, $<\cos^2 \psi>_\psi > 1/3$, i.e., $\psi < 54.7$ deg, $\psi < \pi/2$, and a local maximum when the inverse is true. In the surface averaging case, therefore, local minima and maxima are both represented in the quadrant $0 \le \psi \le \pi/2$, while in the volume average cases the local minimum extends over the whole of the lower quadrant, the local maximum over the whole of the upper quadrant $\pi/2 \le \Psi \le \pi$ (i.e., $\pi/2 \ge \psi' \ge 0$).

In the second example, the cone axis $\underline{C}$ makes the same angle θ with the separation vector as does the transition moment $\underline{E_1}$, i.e., ξ is put equal to θ, while the azimuth Φ of $\underline{C}$ about the $\underline{E_1},\underline{R}$ plane is π. The same construction as in the first example serves to relate θ_c and $\emptyset_c$ to θ, ψ, and γ, and with $\varphi = \pi - \emptyset_c$ the average values for surface distribution (ii,c) are

$$< \kappa^2 > = (1 + \cos^2 \theta)^2 \cos^2 \psi + \frac{1}{2} \cos^2 \theta \sin^2 \theta \sin^2 \psi$$

$$< \cos^2 \theta_T >_d = (1 - 2 \cos^2 \theta)^2 \cos^2 \psi + 2 \cos^2 \theta \sin^2 \theta \sin^2 \psi$$

$$< \kappa^2 \cos^2 \theta_T > = [(1 + \cos^2 \theta)(1 - 2 \cos^2 \theta)]^2 \cos^4 \psi$$

$$+ [\frac{1}{2}(1 + 4 \cos^2 \theta)^2 - 2(1 + \cos^2 \theta)$$

$$\cdot (1 - 2 \cos^2 \theta)] \cos^2 \theta \sin^2 \theta \cos^2 \psi \sin^2 \psi$$

$$+ \frac{3}{2} \cos^4 \theta \sin^4 \theta \sin^4 \psi$$

Figure 49 displays the orientation and depolarization factors derived from these relationships, while those for the volume averaging cases (ii,C) and (ii,C′) appear in Figs. 50 and 51. In these situations it can be shown that the values of $< \kappa^2 >$ represent local maxima and minima and that the condition for the former is that $\cos^2 \psi$, $< \cos^2 \psi >_\psi > 1/3$ and inversely for the latter, i.e., the opposite set of conditions from those of the first example.

Local maxima and minima of $< \kappa^2 >$ may also be found for every value of the azimuth Φ, the condition on them being that

$$\sin \theta \sin \xi \cos \Phi - 2 \cos \theta \cos \xi = 0$$

corresponding again to a zero value of κ for a rigid transition moment along $\underline{C}$.

APPENDIX C: LUMINESCENCE
DEPOLARIZATION BY RAPID, RESTRICTED MOTION

In the first part of this Appendix two simple models which are designed to approximate the rapid motion of a luminophore relative to a macromolecular substrate are discussed and used to evaluate the depolarization factors resulting from such motion. The second part of the Appendix deals with the time dependence of the EA of a luminophore which is free to rotate at an arbitrary rate relative to the macromolecular substrate.

1. Two Models for Luminophore Depolarization

From the considerations of Appendix A, if an excited luminophore moves through an arbitrary angle θ', the average EA is decreased from the stationary isotropic solution value $r_f = 0.4$ for coincident absorption and emission transition moments by

$$<d'> = \frac{3}{2} <\cos^2 \theta'> - \frac{1}{2} \qquad (A27)$$

If the motion is restricted, it may be approximated by various geometrical models for analytical purposes (compare Sec. III), two of the simplest being considered here. The first model assumes rotation about a single bond linking the luminophore to a stationary substrate molecule, the transition moment making an angle θ with this bond. If the motion is rapid (dynamic) with respect to the emission lifetime, and if all available orientations are sampled with equal probability, the relevant averaging is over the surface of a cone of half-angle θ, so that (see Fig. 58)

$$\cos \theta' = \cos^2 \theta + \sin^2 \theta \sin \phi \qquad (A28)$$

By averaging over the azimuthal angle ϕ,

$$<\cos^2 \theta'>_d = \cos^4 \theta + \frac{1}{2} \sin^4 \theta \qquad (A29)$$

or, by rearrangement,

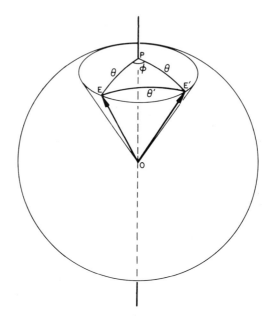

FIG. 58.   Geometry for depolarization over the surface of a cone.

$$<d'>_d = \frac{3}{2}<\cos^2\theta'>_d - \frac{1}{2} = \left(\frac{3}{2}\cos^2\theta - \frac{1}{2}\right)^2 \qquad (A30)$$

If rotation can occur about several bonds linking the luminophore and the substrate, a more suitable model is one in which all transition moment orientations within a cone are equally likely and are sampled during the emission lifetime. Whereas in the first case the average was independent of the luminophore orientation at the moment of excitation, it is readily seen from Fig. 59 that for this model the average has to be taken over all the possible initial orientations. For $0 \le \Theta \le \pi/2$, an emission vector characterized by θ will undergo a reorientation whose extent depends on the limits $\pm\Phi$ of the azimuthal angle ϕ and is given by

$$<\cos^2\theta'>_\theta$$

$$= \frac{\displaystyle\int_{\theta'=0}^{\Theta-\theta}\int_{\phi=0}^{2\pi}\cos^2\theta'\sin\theta'd\theta'd\phi + \int_{\theta'=\Theta-\theta}^{\Theta+\theta}\int_{\phi=-\Phi}^{\Phi}\cos^2\theta'\sin\theta'd\theta'd\phi}{\displaystyle\int_{\theta'=0}^{\Theta-\theta}\int_{\phi=0}^{2\pi}\sin\theta'd\theta'd\phi + \int_{\theta'=\Theta-\theta}^{\Theta+\theta}\int_{\phi=-\Phi}^{\Phi}\sin\theta'd\theta'd\phi}$$

$$(A31)$$

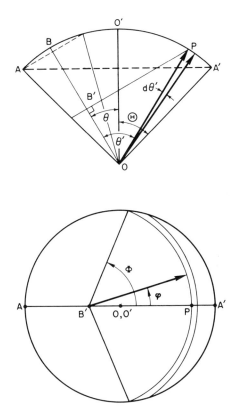

FIG. 59. Geometry for depolarization over the volume of a cone.
Upper: longitudinal section; Lower: plain view.

while by averaging over all possible values of θ corresponding to
different initial orientations,

$$<\cos^2 \theta'>_\Theta = \frac{\int_0^\Theta <\cos^2 \theta'>_\theta \sin \theta \, d\theta}{\int_0^\Theta \sin \theta \, d\theta} \tag{A32}$$

In order to perform the necessary integrations in Eqs. (A31)
and (A32), use is made of the geometrical relationship

$$\cos \Theta = \cos \theta \cos \theta' + \sin \theta \sin \theta' \cos \Phi \tag{A33}$$

obtaining, in the range $0 < \Theta \leq \pi/2$,

$$\langle \cos^2\theta' \rangle_\Theta$$

$$= \left[\frac{1}{1-\cos\Theta}\right] \int_{\theta=0}^{\Theta} \frac{\left[\frac{\pi}{3}\left(1-\cos^3(\Theta-\theta)\right) + \int_{\theta'=\Theta-\theta}^{\Theta+\theta} \arccos\left(\frac{\cos\Theta - \cos\theta\cos\theta'}{\sin\theta\sin\theta'}\right)\cos^2\theta'\sin\theta'\,d\theta'\right]}{\left[\pi\left(1-\cos(\Theta-\theta)\right) + \int_{\theta'=\Theta-\theta}^{\Theta+\theta} \arccos\left(\frac{\cos\Theta - \cos\theta\cos\theta'}{\sin\theta\sin\theta'}\right)\sin\theta'\,d\theta'\right]} \sin\theta\,d\theta$$

(A34a)

Similarly, for the range $\pi/2 \le \Theta < \pi$ corresponding to $\pi/2 \ge \Theta' > 0$,

$$\langle \cos^2\theta' \rangle_{\Theta'}$$

$$= \left[\frac{1}{\cos\Theta'}\right] \int_{\theta=\Theta'}^{\pi/2} \frac{\left[\frac{\pi}{3}\left(1-\cos^3(\theta-\Theta')\right) + \int_{\theta'=\theta-\Theta'}^{\theta+\Theta'} \arccos\left(\frac{\cos\theta\cos\theta' - \cos\Theta'}{\sin\theta\sin\theta'}\right)\cos^2\theta'\sin\theta'\,d\theta' + \frac{\pi}{3}\left(1+\cos^3(\theta+\Theta')\right)\right]}{\left[\pi\left(1-\cos(\theta-\Theta')\right) + \int_{\theta'=\theta-\Theta'}^{\theta+\Theta'} \arccos\left(\frac{\cos\theta\cos\theta' - \cos\Theta'}{\sin\theta\sin\theta'}\right)\sin\theta'\,d\theta' + \pi\left(1+\cos(\theta+\Theta')\right)\right]} \sin\theta\,d\theta$$

(A34b)

The double integrals were evaluated by computer using the Gaussian quadrature method, and the results are displayed in Fig. 52

(Sec. IV) as the depolarization factor: $<d'>_d = 3/2 <\cos^2 \theta'> - 1/2$, the averages being taken over Θ and Θ', along with those for the single bond rotation model. A more simplistic and incorrect model assumes restricted motion over the volume of a cone as above, but a unique direction for the excitation vector identical with the symmetry axis of the cone. The corresponding depolarization factors for such a model are included in Fig. 52 for comparison.

Other volume filling models might also be considered, e.g., with Gaussian or Lorentzian distributions about a mean position, but it seems unlikely that the applications of such models, although physically more realistic, would have any great advantage in practice over the simple geometrical distribution within the volume of a cone described above.

2. Time-Dependent Emission
Anisotropy of Luminophores with Restricted Motion

As pointed out in Sec. IV, the relative magnitudes of macromolecular and segmental motional correlation times determine the time course of depolarization and, along with the emission lifetime, the steady state value of the EA under particular conditions of temperature and solvent viscosity. A short derivation of the time dependence of the EA of a luminophore rotating at an arbitrary rate about a unique axis fixed in an isotropically rotating macromolecule is given below. Equation (A30) represents the limit of this expression, as the independent rotation becomes very rapid in comparison with the macromolecular motion and luminophore emission rate.

Following the method proposed by Steiner and McAlister [73], the time-dependent EA may be written as a product of depolarizing factors describing the independent rotation of the luminophore about an axis fixed in the macromolecular framework and rotations of the axis itself (see Eq. (14) and Appendix A)

$$r(t) = r_f\left(\frac{3}{2}<\cos^2\lambda_m(t)> - \frac{1}{2}\right)\left(\frac{3}{2}<\cos^2\lambda_s(t)> - \frac{1}{2}\right) \qquad (A35)$$

where λ_s is the angle through which the rotating segment bearing the luminophore moves in time t, λ_m that for movement of the axis itself. The average values are those of $\cos^2\lambda_{m,s}$ at time t after excitation. If the absorption and emission vectors $\underline{A}$ and $\underline{E}$ make angles θ_A and θ_E with the axis while the azimuthal angle between them in the plane orthogonal to the axis is ξ, and the two vectors rotate through an angle $\delta(t)$ in this plane between time zero and t (Fig. 60), the change (relative to the axis) in orientation between absorption and emission is given by

$$\cos\lambda_s(t) = \cos\theta_A \cos\theta_E + \sin\theta_A \sin\theta_E \cos[\xi \pm \delta(t)] \quad (A36)$$

By substituting this into Eq. (A35), and after some rearrangement, the time-dependent EA is seen to be

$$r(t) = r_f\left[\frac{3}{2}\cos^2\lambda_m(t) - \frac{1}{2}\right]\left[\left(\frac{3}{2}\cos^2\theta_A - \frac{1}{2}\right)\left(\frac{3}{2}\cos^2\theta_E - \frac{1}{2}\right)\right.$$

$$+ \frac{3}{4}\sin^2\theta_A \sin^2\theta_E\left(2\cos^2\xi - 1\right)\left(2<\cos^2\delta(t)> - 1\right)$$

$$\left. + 3\cos\theta_A \sin\theta_A \cos\theta_E \sin\theta_E \cos\xi <\cos\delta(t)>\right] \qquad (A37)$$

For rotation about a single axis, the averages have been given by Weber [38] as

$$<\cos\delta(t)> = \exp(-t/2\rho_s)$$

$$2<\cos^2\delta(t)> - 1 = \exp(-2t/\rho_s) \qquad\qquad (A38)$$

where ρ_s is the relaxation time of the segmental rotation which is related to the rotational correlation time τ_s by $\rho_s = 3\tau_s$.

For isotropic rotation of the axis fixed in the macromolecular frame, the equivalent average is given analogously [38] by

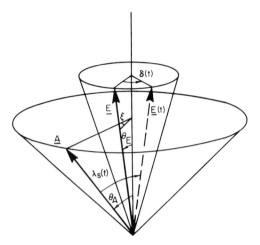

FIG. 60. Geometry for time-dependent depolarization due to rotation of absorption and emission transition moment vectors about a
fixed axis.

$$\frac{3}{2} < \cos^2 \lambda_m(t)> - \frac{1}{2} = \exp[-3t/\rho_m] \tag{A39}$$

In this case, then, the time-dependent EA is, according to Eq. (A37),
described by three exponentials as indicated in Sec. IV [74]:

$$r(t) = r_f[Ae^{-3t/\rho_m} + Be^{-(3/\rho_m + 2/\rho_s)t} + Ce^{-(3/\rho_m + 1/2\rho_s)t}] \tag{A40}$$

The equivalent expression for the average EA is readily obtained from Eq. (A40) by averaging over the decay rate $e^{-t/\tau}$ when

$$<r> = r_f\left[\frac{A}{1 + \dfrac{3\tau}{\rho_m}} + \frac{B}{1 + \dfrac{3\tau}{\rho_m} + \dfrac{2\tau}{\rho_s}} + \frac{C}{1 + \dfrac{3\tau}{\rho_m} + \dfrac{\tau}{2\rho_s}} \right] \tag{A41}$$

which is the result obtained by Gottlieb and Wahl [42].

If ρ_s is very much smaller than ρ_m and τ, only the first terms
in Eqs. (A40) and (A41) will be observed experimentally, and it is
readily seen that the pre-exponential factor A corresponds to $<d'>_d$

of Eq. (A30) if, in addition, the absorption and emission moments
coincide $(\theta_1 = \theta_2 = \theta)$, i.e., $r_f A = r_{om}$. Similarly, when ρ_s is
merely somewhat smaller than ρ_m, the first term will predominate at
long times.

3. Time-Dependent Emission
Anisotropy of Rigidly Attached Luminophores

For absorption and emission moments <u>fixed</u> with respect to the
macromolecular framework, solutions have been obtained for the
emission anisotropy decay of an ellipsoid of revolution by Memming
[75] and more recently by three independent groups [76-78] for that
of a general ellipsoid, which relate the directions of the moments
to the principal axes of the ellipsoid. The ellipsoids are supposed
to correspond to the hydrodynamic entities comprising the lumino-
phore-bearing molecule or macromolecule together with any associated
solvation shell. The expressions for anisotropy decay contain three
and five exponentials, respectively, for these two cases. The rotation of an axis fixed in the macromolecule is also described by
the sum of three or five exponentials,

$$\frac{3}{2}<\cos^2 \lambda_m(t)> - \frac{1}{2} = \sum_{i=1}^{3,5} A_i e^{-t/\tau_i} , \qquad (A42)$$

and the time-dependent EA for rotation about this axis is thus by
analogy with Eq. (A40) given in general by the sum of nine or fifteen
exponentials. A different approach to the problem of anisotropic
rotations [79,80] leads to a reduction in the number of exponentials
operationally describing the time-dependent EA which may simplify
the extension to the case of segmental motions. If ρ_s is very small
compared with τ_i and τ, then again $<d'>_d$ is obtained from the value
of the EA at zero time by use of Eq. (31). Obviously, it will be
difficult if not impossible to obtain a value for $<d'>_d$ when ρ_s is
not very different from τ_i and/or τ.

The fundamental EA which appears in all these equations has the value 0.4 or 0.2 for polarized or natural excitation, respectively. In using, e.g., Eq. (A40) or its time-average equivalent (A41), however, it is tacitly assumed that r_f takes on the value $r(0)$ or r_o, the observable limiting value which is almost invariably less than r_f (see discussion in Sec. IV). If this difference is attributable to some ultrafast depolarization process, e.g., torsional oscillations [50], or to the presence of a three-dimensional oscillator [51], then the effect of this might be represented in Eq. (A35) by a third depolarization factor of similar form to that describing rotation about a single axis. Since the correlation time for this process is ultrashort, only a constant multiplier of r_f would be experimentally evident, at least in the nanosecond range of resolution, even when the solutions are highly viscous. It is, therefore, justifiable to use Eqs. (A40) and (A41) and the analogous equations for anisotropy decay of ellipsoids with rigidly attached luminophores [75-80], wherein r_f is replaced by r_o. A detailed proof has been given by Wahl and coworkers [81].

ADDENDUM

Since submission of this manuscript to the publishers, part of the work detailed therein, together with some newer results has been presented elsewhere [82,83] (see also [62]). In addition, several new determinations of intramolecular separations in macromolecules have appeared in the literature during this time, utilizing both native [84-86] and specifically conjugated [86-89] fluorophores. In particular, the topographies of the formyl-methionine and glutamate tRNA's of Escherichia coli have been examined in some detail with several D-A pairs [89]. Although variable rotational correlation times for the singly-labeled tRNA's were reported (6.5 - 24 nsec), indicating the presence of segmental flexibility for some of the labels, no zero point anisotropies (r_o) were given so that here, as

in the other cases, estimates of maximum and minimum possible derived separations cannot be made. However, most of the distances
calculated assuming $<\kappa^2> = 2/3$ were close to those calculated from
x-ray crystallographic data on yeast $tRNA^{Phe}$.

Finally, it is pertinent to record that two excellent compre-
hensive reviews on the technique and theory of time-dependent polarized emission spectroscopy and its application to macromolecular
systems have appeared recently [90,91].

REFERENCES

1. Th. Förster, Ann. Phys., 2, 55 (1948).

2. Th. Förster, in Modern Quantum Chemistry (O. Sinanoğlu, ed.),
    Part III, Academic Press, New York, 1965, p. 93.

3. K. Beardsley and C. R. Cantor, Proc. Natl. Acad. Sci., 65, 39
    (1970).

4. J. Eisinger, Biochemistry, 8, 3902 (1969).

5. S. A. Latt, D. S. Auld, and B. L. Vallee, Proc. Natl. Acad. Sci.,
    67, 1383 (1970).

6. T. C. Werner, J. R. Bunting, and R. E. Cathou, Proc. Natl. Acad.
    Sci., 69, 795 (1972).

7. P. W. Schiller, Proc. Natl. Acad. Sci., 69, 975 (1972).

8. C.-W. Wu and L. Stryer, Proc. Natl. Acad. Sci., 69, 1104 (1972).

9. L. Stryer and R. P. Haugland, Proc. Natl. Acad. Sci., 58, 719
    (1967).

10. J. Eisinger, B. Feuer, and A. A. Lamola, Biochemistry, 8, 3908
    (1969).

11. S. A. Latt, H. T. Cheung, and E. R. Blout, J. Amer. Chem. Soc.,
    87, 995 (1965).

12. G. Gabor, Biopolymers, 6, 809 (1968).

13. G. Weber and F. W. J. Teale, Disc. Faraday Soc., 27, 134 (1959).

14. R. P. Haugland, J. Yguerabide, and L. Stryer, Proc. Natl. Acad. Sci., 63, 23 (1969).

15. L. Stryer, Science, 162, 526 (1968).

16. G. K. Radda, in Current Topics in Bioenergetics (D. R. Sanadi, ed.), Vol. 4, Academic Press, New York, 1971, p. 81.

17. I. Z. Steinberg, Ann. Rev. Biochem., 40, 83 (1971).

18. M. D. Galanin, Soviet Phys.-JETP, 1, 317 (1955).

19. M. Z. Maksimov and I. M. Rozman, Opt. Spectry., 12, 337 (1962).

20. I. Z. Steinberg, J. Chem. Phys., 48, 2411 (1968).

21. C. Ka Luk, Biopolymers, 10, 1317 (1971).

22. P. P. Feofilov, The Physical Basis of Polarized Emission, Consultants Bureau, New York, 1961.

23. A. Jabłoński, Acta Phys. Polon., 16, 471 (1957).

24. A. Jabłoński, Bull. Acad. Polon. Sci., Ser. Sci. Math. Astr. Phys., 8, 259 (1960).

25. G. Weber and S. R. Anderson, Biochemistry, 8, 361 (1969).

26. S. R. Anderson and G. Weber, Biochemistry, 8, 371 (1969).

27. R. E. Dale and F. W. J. Teale, Photochem. Photobiol., 12, 99 (1970).

28. D. Genest, Ph. Wahl, and J.-C. Auchet, Biochim. Biophys. Acta, 259, 175 (1972).

29. Ph. Wahl, Chapter 1 of this volume.

30. R. A. Badley and F. W. J. Teale, J. Mol. Biol., 58, 567 (1971).

31. R. B. Gennis and C. R. Cantor, Biochemistry, 11, 2509 (1972).

32. L. S. Gennis, R. B. Gennis, and C. R. Cantor, Biochemistry, 11, 2517 (1972).

33. A. Jabłoński, Z. Naturforsch., 16a, 1 (1961).

34. T. Tao, Biopolymers, 8, 609 (1969).

35. Ph. Wahl, M. Kasai, J.-P. Changeux, and J.-C. Auchet, Eur. J. Biochem., 18, 332 (1971).

36. R. Schuyler and I. Isenberg, Rev. Sci. Instr., 42, 813 (1971).

37. I. Isenberg, Chapter 2 of this volume.

38. G. Weber, Adv. Protein Chem., 8, 415 (1953).

39. F. Perrin, J. Phys. Radium, 7, 390 (1926).

40. J. Yguerabide, H. F. Epstein, and L. Stryer, J. Mol. Biol., 51,
     573 (1970).

41. J.-C. Brochon, Ph. Wahl, and J.-C. Auchet, Eur. J. Biochem.,
     25, 20 (1972).

42. Y. Ya Gottlieb and Ph. Wahl, J. Chim. Phys., 60, 849 (1963).

43. Ph. Wahl and G. Weber, J. Mol. Biol., 30, 371 (1967).

44. D. Wallach, J. Chem. Phys., 47, 5258 (1967).

45. G. Weber, J. Opt. Soc. Am., 46, 962 (1956).

46. G. Weber, Biochem. J., 51, 145 (1952).

47. R. E. Dale and R. K. Bauer, Acta Phys. Polon., A40, 853 (1971).

48. F. W. J. Teale, Photochem. Photobiol., 10, 363 (1969).

49. G. Weber, Biochem. J., 75, 335 (1960).

50. A. Jabłoński, Acta Phys. Polon, 10, 193 (1950).

51. A. Jabłoński, Z. Phys., 96, 236 (1935).

52. Y. Nishijima, in Luminescence of Organic and Inorganic Materials
     (H. P. Kallmann and G. Marmor Spruch, eds.), Wiley, New York,
     1962, p. 235.

53. S. S. Rathi and M. K. Machwe, Phys. Lett., 25A, 41 (1967).

54. A. Jabłoński, Z. Phys., 103, 526 (1936).

55. R. D. Spencer and G. Weber, J. Chem. Phys., 52, 1654 (1970).

56. R. A. Badley and F. W. J. Teale, J. Mol. Biol., 44, 71 (1969).

57. R. F. Chen, H. Edelhoch, and R. F. Steiner, in Physical Prin-
     ciples and Techniques of Protein Chemistry (S. J. Leach, ed.),
     Academic Press, New York, 1969, p. 171.

58. L. Brand and B. Witholt, in Methods in Enzymology (C. H. W.
     Hirs, ed.), Vol. XI, Academic Press, New York, 1967, p. 776.

59. R. D. Rauh, T. R. Evans, and P. A. Leermakers, J. Amer. Chem. Soc., 91, 1868 (1969).

60. K. Beardsley, T. Tao, and C. R. Cantor, Biochemistry, 9, 3524 (1970).

61. T. Tao, J. H. Nelson, and C. R. Cantor, Biochemistry, 9, 3514 (1970).

62. W. E. Blumberg, R. E. Dale, J. Eisinger, and D. M. Zuckerman, Biopolymers, 13, 1607 (1974).

63. J. E. Churchich, Biochim. Biophys. Acta, 75, 274 (1963).

64. S. H. Kim, G. J. Quigley, F. L. Suddath, A. McPherson, D. Sneden, J. J. Kim, J. Weinzierl, and A. Rich, Science, 179, 285 (1973).

65. C.-G. Jang, Biochem. Biophys. Res. Commun., 50, 612 (1973).

66. C. R. Cantor and P. Pechukas, Proc. Natl. Acad Sci., 68, 2099 (1971).

67. A. Grinvald, E. Haas, and I. Z. Steinberg, Proc. Natl. Acad. Sci., 69, 2273 (1972).

68. F. Perrin, Ann. Phys., 12, 169 (1929).

69. P. Soleillet, Ann. Phys., 12, 23 (1929).

70. G. Weber, in Fluorescence and Phosphorescence Analysis (D. M. Hercules, ed.), Wiley-Interscience, New York, 1966, p. 217.

71. Fluorescence Spectroscopy. An Introduction for Biology and Medicine (A. J. Pesce, C.-G. Rosén, and T. L. Pasby, eds.), Marcel Dekker, Inc., New York, 1971.

72. A. G. Tweet, W. D. Bellamy, and G. L. Gaines, Jr., J. Chem. Phys., 41, 2068 (1964).

73. R. F. Steiner and A. J. McAlister, J. Polymer Sci., 24, 105 (1957).

74. Ph. Wahl, C. R. Acad. Sci., 260, 6891 (1965).

75. R. Memming, Z. Phys. Chem., 28, 168 (1961).

76. T. J. Chuang and K. B. Eisenthal, J. Chem. Phys., 57, 5094 (1972).

77. M. Ehrenberg and R. Rigler, Chem. Phys. Lett., 14, 539 (1972).

78. G. G. Belford, R. L. Belford, and G. Weber, Proc. Natl. Acad. Sci., 69, 1392 (1972).

79. G. Weber, J. Chem. Phys., 55, 2399 (1971).

80. G. Weber, Ann. Rev. Biophys. Bioeng., 1, 553 (1972).

81. Ph. Wahl, G. Meyer, J. Parrod, and J.-C. Auchet, Eur. Polymer J., 6, 585 (1970).

82. J. Eisinger and R. E. Dale, J. Mol. Biol., 84, 643 (1974).

83. R. E. Dale and J. Eisinger, Biopolymers, 13, 1573 (1974).

84. R. M. Epand, Photochem. Photobiol., 18, 245 (1973).

85. M. Takahashi and R. A. Harvey, Biochemistry, 12, 4743 (1973).

86. O. A. Moe, Jr., D. A. Lerner, and G. Hammes, Biochemistry, 13, 2552 (1974).

87. J. R. Bunting and R. E. Cathou, J. Mol. Biol., 77, 223 (1973).

88. J. R. Bunting and R. E. Cathou, J. Mol. Biol., 87, 329 (1974).

89. C.-H. Yang and D. Söll, Proc. Natl. Acad. Sci., 71, 2838 (1974).

90. J. Yguerabide, Nanosecond Fluorescence Spectroscopy of Macromolecules, in Methods in Enzymology (C. H. W. Hirs and S. N. Timasheff, eds.), Vol. XXVI, Part C, Academic Press, New York, 1972, p. 498.

91. R. Rigler and M. Ehrenberg, Quart. Rev. Biophys., 6, 139 (1973).

Chapter 5

THE MEASUREMENT OF
INTRAMOLECULAR DISTANCES BY ENERGY TRANSFER

Peter W. Schiller*
McCollum-Pratt Institute
The Johns Hopkins University
Baltimore, Maryland

I.   INTRODUCTION

The use of long-range nonradiative transfer of electronic excitation energy for the measurement of intramolecular distances in
biopolymers has attracted much attention in the past decade. Some

*Present address: Clinical Research Institute of Montreal, Montreal,
Quebec, Canada.

considerations pertinent to energy transfer in proteins and polypeptides as well as numerous examples have recently been presented
in an excellent review by Steinberg [1].

This contribution aims to focus on some methodical aspects of
the measurement of intramolecular distances which will be illustrated
with model compounds as well as with a dansylated derivative of the
adrenocorticotropic hormone (ACTH).

The difficulty of obtaining the value of the orientation factor
needed for the computation of the distance by Förster's equation has
generally been recognized as a major weakness of the method and is
dealt with elsewhere in this volume [2]. The observation [3] that
the presence of an acceptor chromophore sometimes introduces additional quenching of the donor fluorescence aside from energy transfer by the dipole-dipole mechanism revealed another potential pit-
fall of crucial importance which will be one of the major points
emphasized in this chapter. The occurrence of such additional
quenching does not exclude the determination of the intramolecular
distance, but if it is not recognized, significant errors in the
calculated distance may result. The presence or absence of such
additional quenching can be verified by monitoring both the relative
decrease of the donor fluorescence and the relative increase of the
acceptor fluorescence. Therefore, in an ideal energy transfer experiment, both donor and acceptor should be fluorescent, and their
emission spectra should be well separated. This situation is at-
tainable and is assumed in the following discussion.

II. THE DETERMINATION OF
TRANSFER EFFICIENCY BY VARIOUS METHODS

The rates of the various processes affecting the first excited
singlet state of a donor (D) and an acceptor (A) are depicted in
Fig. 1 in a simplified fashion which allows the definition of all

the experimental parameters used for the quantitative determination of the efficiency of transfer by various methods. k_{FD} is the decay rate of fluorescence (the reciprocal of the natural lifetime) and k_{QD} the sum of all other quenching processes in the donor (internal conversion plus external conversion processes). The process of energy transfer occurring at the rate k_T competes with these processes more or less efficiently in the depopulation of the first excited state of the donor, and the efficiency of transfer E is therefore defined as

$$E = \frac{k_T}{k_{FD} + k_{QD} + k_T} \tag{1}$$

k_{FA} and k_{QA} describe the rates of natural fluorescence and the sum of all other quenching processes, respectively, in the acceptor.

The transfer efficiency can be determined from measurements of quantum yields of either donor or acceptor by means of the steady-state technique. Alternatively, it can also be obtained from lifetime measurements of the donor and, in principle, also by analysis of the decay of acceptor fluorescence which is best accomplished by

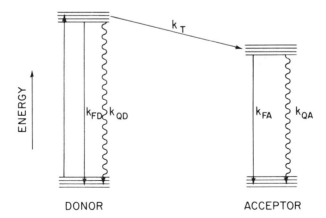

FIG. 1. Rates of processes affecting excited states of donor and acceptor.

the time-correlated single photon counting technique. There are
thus four methods available for the quantitative evaluation of intramolecular energy transfer.

A. Measurements of Quantum Yields or Lifetimes of the Donor

The determination of E from either quantum yields or lifetimes
of the donor in the presence and absence of transfer is very
straightforward and is outlined in the following scheme:

Steady-state technique	Lifetime measurement
Quantum yield of donor (Eq.)	Fluorescence lifetime of donor (Eq.)

Absence of transfer:

$$\varphi_D^{\,\circ} = \frac{k_{FD}}{k_{FD} + k_{QD}} \quad (2) \qquad \tau_D^{\,\circ} = \frac{1}{k_{FD} + k_{QD}} \quad (4)$$

Presence of transfer:

$$\varphi_D = \frac{k_{FD}}{k_{FD} + k_{QD} + k_T} \quad (3) \qquad \tau_D = \frac{1}{k_{FD} + k_{QD} + k_T} \quad (5)$$

$$E = 1 - \frac{\varphi_D}{\varphi_D^{\,\circ}} \quad (6) \qquad E = 1 - \frac{\tau_D}{\tau_D^{\,\circ}} \quad (7)$$

Equations (6) and (7) are simply obtained by combination of Eq. (1)
with Eqs. (2) and (3), and (5) and (6), respectively. The advantage
of using lifetimes rather than quantum yields for the determination
of transfer efficiencies is that the former are measured with greater
accuracy. This accuracy is especially advantageous in cases where
the transfer efficiency is only a few percent or close to 100%. In
such cases it may still be possible to calculate accurate distances
on the basis of measured lifetimes, while the determination of
quantum yields would only permit the statement of an upper or lower
limit for the distance. A good coincidence between transfer effi-

ciencies determined from lifetime measurements and from quantum yields has been reported [4].

The usefulness of complementing lifetime measurements with quantum yield determinations has recently been demonstrated in a case [6] where incomplete labeling of a donor-containing macromolecule with an acceptor could have provided a misleading result on the basis of lifetime measurements alone. In this particular case, the unchanged donor lifetime of the labeled sample was due to a small fraction of the macromolecules which did not contain an acceptor, while steady-state measurements revealed that the transfer in the molecules containing both donor and acceptor was 100% efficient.

It is important that the measurement of the donor quantum yield, $\varphi_D^{\,\circ}$, or lifetime, $\tau_D^{\,\circ}$, in the absence of transfer be performed with the donor attached to the same site of the macromolecule which it occupies in the actual energy transfer experiment. This is absolutely required, since quantum yields and lifetimes of fluorophores in biopolymers are extremely dependent on the microenvironment (polarity, presence of a charged group, etc.). The correct measurement of $\varphi_D^{\,\circ}$ or $\tau_D^{\,\circ}$ has either been achieved with the macromolecule containing the donor but devoid of the acceptor [3 to 7b], or by bleaching of the acceptor [8].

The determination of the transfer efficiency from the donor fluorescence by either method requires a good separation of the donor and acceptor emission spectra. Greatly overlapping emission spectra make subtraction procedures necessary which may decrease the accuracy of the determination significantly. Both methods by themselves do not establish the actual occurrence of resonance energy transfer. They do not verify whether the amount of quenching observed is exclusively due to energy transfer or whether it is partially or even fully caused by other quenching processes resulting from the presence of the acceptor. Complementary studies on the basis of the acceptor fluorescence are therefore encouraged.

B. Increase of Acceptor Fluorescence

The determination of the transfer efficiency from the relative
increase of the acceptor fluorescence is somewhat more complicated,
mainly due to the fact that with most donor-acceptor pairs some
direct excitation of the acceptor occurs upon irradiation in the
donor absorption range. The acceptor thus emits both direct and
sensitized fluorescence. In the rare case where the acceptor does
not absorb in the excitation range of the donor, the transfer efficiency may be calculated [9] by means of Eq. (8):

$$E = \frac{A_A(\lambda_A)I(\lambda_D)}{A_D(\lambda_D)I_A(\lambda_A)} \tag{8}$$

In the more common case, where direct excitation of the acceptor
does occur, Eq. (9) applies [2]:

$$E = \frac{A_A(\lambda_D)}{A_D(\lambda_D)} \left(\frac{I(\lambda_D)}{I_A(\lambda_D)} - 1 \right) \tag{9}$$

where $A_D(\lambda)$, $A_A(\lambda)$ = absorbance at wavelength λ of donor and acceptor, respectively; $I(\lambda)$, $I_A(\lambda)$ = intensities of acceptor emission
in presence and absence of donor (with excitation at wavelength λ);
and λ_D, λ_A = wavelengths in absorption ranges of donor and acceptor,
respectively. It should be noted that both Eqs. (8) and (9) apply
only to solutions with low absorption. A modified version of Eq.
(8) which permits the determination of E in solutions with high absorption has been given [10] and used in a case where direct excitation of the acceptor was assumed to be absent. Equation (9) is
somewhat simpler than that used by Conrad and Brand [3] which is
based on the definition of the acceptor quantum yield in the absence
of transfer at the wavelength of the acceptor absorption maximum
rather than at the excitation wavelength used in the actual energy
transfer experiment.

In the case of simultaneous, direct excitation of the acceptor,
it is important that the ratio of the absorbance of the donor to

that of the acceptor at the excitation wavelength is as large as
possible.

C. Analysis of the Acceptor Fluorescence Decay

The rate equation that describes the decay of the acceptor
fluorescence in the presence of transfer can be written as

$$\frac{d[A^*]}{dt} = k_T[D^*] - (k_{FA} + k_{QA})[A^*] \tag{10}$$

where $[A^*]$ and $[D^*]$ are the concentrations of excited acceptor and
donor, respectively. The solution of this differential equation
for the case where there is no direct excitation of the acceptor
yields

$$[A^*] = \frac{k_T[D^*]_0}{1/\tau_A^o - 1/\tau_D} (e^{-t/\tau_D} - e^{-t/\tau_A^o}) \tag{11}$$

where $[D^*]_0$ is the concentration of excited donors at the instant of
excitation. The decay of the sensitized acceptor fluorescence is
obviously governed by a sum of two exponentials whose amplitude factors are equal in magnitude but opposite in sign.

In the case where direct excitation of the acceptor has to be
taken into account, the solution of Eq. (10) leads to

$$[A^*] = \frac{k_T[D^*]_0}{1/\tau_A^o - 1/\tau_D} e^{-t/\tau_D} + ([A^*]_0 - \frac{k_T[D^*]_0}{1/\tau_A^o - 1/\tau_D}) e^{-t/\tau_A^o} \tag{12}$$

where $[A^*]_0$ is the concentration of directly excited acceptor molecules at the instant of excitation (time 0). Equations (11) and
(12) are of the same form as the equations given by Birks [11] for
the case of <u>intermolecular</u> transfer under the assumption of Stern-
Volmer kinetics.

From this equation, it becomes evident that direct excitation
of the acceptor affects only one amplitude factor and has no influence on the exponentials. Obviously, the experimental decay curve

of the acceptor requires analysis of two exponential components by
an appropriate method, such as the method of moments [12]. From
such an analysis, τ_D should be obtainable for the computation of the
transfer efficiency by means of Eq. (7). No determinations of the
efficiency of intramolecular energy transfer by analysis of the acceptor decay have been reported so far. It can be predicted that
such analysis will be limited to distances not too far from the
Förster critical distance, especially in the common case where direct excitation of the acceptor also occurs. If the transfer efficiency is only a few percent, the amplitude factor of the first term
in Eq. (12) is small compared to that of the second term, and an
accurate two-component analysis may become difficult. On the other
hand, if the transfer efficiency is close to 100%, the contribution
of the first term in Eq. (12) goes rapidly to zero, and the decay
would essentially be governed by the single exponential, $1/\tau_A^{\,0}$. In
this case, it is impossible to establish on the basis of the decay
constant alone whether the transfer efficiency is 0 or 100%.

An additional difficulty may arise in case the fluorescence
decay of the acceptor attached to the macromolecule in the absence
of transfer is for some reason not governed by a single exponential
but is more complex as sometimes observed with fluorophores covalently bound to biopolymers. In this case, it should become very
difficult, if not impossible, to extract τ_D by means of the analytical methods available. These various disadvantages suggest that
the analysis of the acceptor fluorescence decay may not be the method
of choice for the determination of the transfer efficiency.

III. CALCULATION OF THE INTRAMOLECULAR DISTANCE

The rate of energy transfer k_T is related to geometrical and
spectroscopic parameters by the well-known Förster equation [Eq.
(13)]:

$$k_T = \frac{9(\ln 10)\kappa^2 J_{AD}}{128 \, \pi^5 N n^4 \tau_e r^6} \tag{13}$$

where κ = dipole-dipole orientation factor, N = Avogadro's number,
n = refractive index, τ_e = natural lifetime of donor fluorescence
($= 1/k_{FD}$), r = donor-acceptor distance, and $J_{AD} = \int_0^\infty F_D(\lambda)\epsilon_A(\lambda)\lambda^4 \, d\lambda$
= spectral overlap integral between the molar decadic absorption
coefficient of the acceptor (ϵ_A) and the spectral distribution of
the fluorescence of the donor, normalized to unity (F_D), modified
by the wavelength factor λ^4. While in principle r could be computed
from this equation, it has become customary to calculate the so-
called F̈orster critical distance, R_o, which is the distance where
deactivation of the donor by means of energy transfer and deactivation by all other quenching mechanisms are of equal probability:

$$k_T = k_{FD} + k_{QD} \tag{14}$$

and which is obtained by combination of Eqs. (2), (13), and (14)
after introduction of the numerical values for the constants as

$$R_o = [8.79 \cdot 10^{-28} \frac{\kappa^2}{n^4} \varphi_D^o J_{AD}]^{1/6} \tag{15}$$

The experimental determination of the various parameters in this relation has been discussed elsewhere [1]. The orientation factor, κ^2,
is dealt with elsewhere in this book [2], and the determination of
φ_D^o has been discussed above. Comparison of Eqs. (13) and (15)
shows that R_o and E are related to r by

$$r = (E^{-1} - 1)^{1/6} R_o \tag{16}$$

which allows the computation of the intramolecular distance.

IV.   ADDITIONAL QUENCHING
INTRODUCED BY THE PRESENCE OF THE ACCEPTOR

The observation that not all of the quenching of the donor fluorescence could be accounted for by energy transfer to the acceptor

as evaluated from the relative increase of the acceptor fluorescence
was made by Conrad and Brand [3]. Such additional quenching seems
to occur at relatively small donor-acceptor separations, and its
cause has not been resolved as yet. However, it is known that at
short distances (< 10 Å) higher multipole and exchange interactions
become significant [14]. It is also conceivable that the introduction of the acceptor causes a conformational change of the macromolecule which might alter the environment of the donor. If such
additional quenching does occur, the evaluation of the transfer
efficiency on the basis of quantum yields or lifetimes of the donor
(method 1) results in too large a value for E, and the computed distance will be too short. On the other hand, an evaluation of the
relative increase in acceptor fluorescence (method 2) does provide
the correct transfer efficiency; however, the actual quantum yield
of the donor is decreased owing to additional quenching. As a consequence, the value of R_o calculated on the basis of a donor quantum
yield which does not take into account the additional quenching will
be too large. This obviously leads to the computation of too large
a distance on the basis of Eq. (16). However, the occurrence of
additional quenching does not necessarily exclude the determination
of the correct distance. The procedure to be applied in such case
is the following (method 3): (i) Determine E from the relative increase of the acceptor fluorescence. (ii) Measure the fluorescence
quantum yield of the donor, φ_D, in presence of the acceptor. (iii)
Use E and φ_D in Eq. (6) and calculate $\varphi_D^{o\prime}$ which is φ_D^o modified by
additional quenching. (iv) Calculate new R_o with $\varphi_D^{o\prime}$ and use it in
Eq. (16) for the computation of the correct distance. This proce-
dure is identical with the one used by Conrad and Brand [3] on the
basis of Eq. (17) which is obtained by combination of Eqs. (6), (15),
and (16):

$$r = [8.79 \cdot 10^{-28} \; \frac{\kappa^2}{n^4} \; \varphi_D \; \frac{1}{E} \; J_{AD}]^{1/6} \tag{17}$$

This procedure again requires sufficient separation of the donor and
acceptor emission spectra both for the determination of the transfer

efficiency from the relative increase of the acceptor fluorescence and for the measurement of the donor quantum yield in the presence of the acceptor.

The effect of additional quenching can be illustrated with the simple model compound N^α-Dns-tryptophan, where energy transfer is expected to occur from the tryptophan donor to the dansyl acceptor. [Dns (=dansyl), 1-dimethylaminonaphthalene-5-sulfonate.] In the N^α-Dns-tryptophan, tryptophan fluorescence is barely detectable, and φ_D can be determined approximately as 10^{-4}. A transfer efficiency of 0.999 is calculated from these values. (The quantum yield determinations were based on a value of 0.14 for the quantum yield of tryptophan in aqueous solution.) However, from the relative increase of the dansyl fluorescence, a transfer efficiency of only 0.53 is obtained. The occurrence of additional quenching is thus clearly established. The effect of the additional quenching on the computation of the distance by the various methods described above is presented in Table 1. Also included in Table 1 are the distances computed from data obtained by Conrad and Brand [3] with the model compound Dns-NH(CH$_2$)$_5$(C=O)-Trp, which is also subject to additional quenching. As predicted, the distance obtained by method 3 lies between those obtained by methods 1 and 2. The discrepancies between the distance values obtained by method 2 and the correct ones resulting from method 3 are especially noteworthy.

TABLE 1

Determination of Intramolecular Distances in Model Compounds

Procedure	Distance (Å)	
	N^α-Dns-L-Trp	Dns-NH(CH$_2$)$_5$(C=O)-Trp
Method 1	6.1	9.8
Method 2	20.5	23.8
Method 3	6.9	11.6

In view of the possibility of additional quenching, it is rec-
ommended that the evaluation of the transfer efficiency be performed
from both quantum yields or lifetimes of the donor and the relative
increase of the acceptor fluorescence. Two cases have been reported
in the literature [7,10] where both procedures yielded the same result; it can therefore be concluded that energy transfer was the
only quenching effect caused by the presence of the acceptor.

V. THE EVALUATION OF ENERGY
TRANSFER IN A DANSYLATED DERIVATIVE OF ACTH

The specific incorporation of a dansyl group in the $ACTH_{1-24}$-
fragment has been achieved by peptide synthesis [15]. The deriva-
tive, N^ϵ-Dansyllysine^{21}-ACTH-(1-24)-tetrakosipeptide ([Lys(Dns)^{21}]-
$ACTH_{1-24}$), contains Trp in position 9 as a donor and Lys(Dns) in
position 21 as an acceptor:

Ser-Tyr-Ser-Met-Glu-His-Phe-Arg-Trp-Gly-Lys-Pro-Val-Gly-
1

Lys-Lys-Arg-Arg-Pro-Val-Lys-Val-Tyr-Pro,
                                        24

|
NH
|
SO$_2$

$(CH_3)_2N-$

The compound displays biological activity similar to that observed
with the native $ACTH_{1-24}$-fragment [15]. The reference compound,
N^ϵ-Dansyllysine^{21}-ACTH-(11-24)-tetradekapeptide ([Lys(Dns)^{21}]ACTH$_{11}$-
$_{24}$), devoid of the donor Trp^9, was also prepared. Figure 2 shows
the excitation spectra of Dns-emission at 570 nm of [Lys(Dns)^{21}]-
$ACTH_{1-24}$, [Lys(Dns)^{21}]ACTH$_{11-24}$, and N^ϵ-Dns-lysine [Lys(Dns)], and

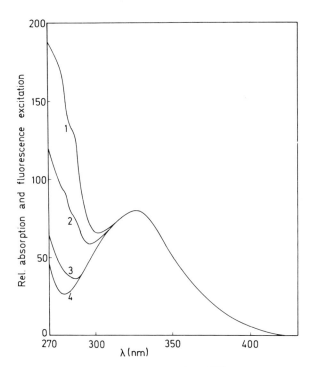

FIG. 2.   Absorption spectra of $[Lys(Dns)^{21}]ACTH_{1-24}$ (curve 1) and Lys(Dns) (curve 4).   Fluorescence excitation spectra of $[Lys(Dns)^{21}-ACTH_{1-24}$ (curve 2), $[Lys(Dns)^{21}]ACTH_{11-24}$ (curve 3) and Lys(Dns) (curve 4).   The emission was measured at 570 nm; the spectra are normalized at 327 nm.

the absorption spectra of $[Lys(Dns)^{21}]ACTH_{1-24}$ and Lys(Dns).   The spectra were corrected and normalized at 327 nm, the long wavelength peak of dansyl absorption in H_2O.   The excitation spectrum of Lys(Dns) is exactly superimposable on its absorption spectrum (curve 4) as expected for an isolated fluorophore, provided that the fluorescence quantum yield is independent of the excitation wavelength.   The occurrence of energy transfer from Trp^9 to $Lys(Dns)^{21}$ is established by inspection of the Dns-fluorescence excitation spectrum of $[Lys(Dns)^{21}]-ACTH_{1-24}$ (curve 2) which shows additional excitation in the wavelength range of tryptophan absorption.   Since energy transfer from Tyr^{23} to $Lys(Dns)^{21}$ also occurs (curve 3), the study of energy transfer from

Trp[9] to Lys(Dns)[21] had to be performed with excitation at wavelengths
above 293 nm, where tyrosine no longer absorbs. Excitation at 293 nm
was used for the determination of the transfer efficiency from the
tryptophan quantum yields and from the relative increase of the Dns-
fluorescence [7]. For the measurement of the donor lifetimes by
means of the monophoton technique, a narrow bandpass filter with
transmittance in the range from 293 to 297 nm was used [16]. The
measurement of the donor quantum yield and lifetime in the absence
of the acceptor could easily be performed with the native fragment
ACTH$_{1-24}$. In aqueous solution the values of φ_D^O and φ_D were 0.100
and 0.060, respectively, and τ_D^O and τ_D were 2.60 nsec and 1.41 nsec,
respectively, as determined with a one-component analysis by means
of the method of moments. The use of these values in Eqs. (6) and
(7), respectively, and an analysis of the acceptor fluorescence
according to Eq. (9) provided the values for the efficiency of
transfer from Trp[9] to Lys(Dns)[21] seen in Table 2. The good agreement between these values demonstrates that in this case energy
transfer is the only quenching process introduced by the presence
of the acceptor. This result should be compared with those obtained
from the model compounds discussed above which contained the same
donor-acceptor pair. Evidently, the phenomenon of additional quenching is limited to relatively short donor-acceptor separations.

TABLE 2

The Efficiency of Transfer Obtained by Various Methods

Method	Transfer efficiency
Quantum yields of the donor	0.40
Lifetimes of the donor	0.46
Relative increase of acceptor fluorescence	0.45

A determination of E from the decay of the acceptor fluorescence failed owing to the fact that the decay of the Dns-fluorescence in the absence of transfer is not governed by a single exponential [16]. It was thus not possible to obtain τ_D from a two-component analysis.

The determination of the other parameters necessary for the computation of the distance has been described elsewhere [7]. An average value of 2/3 was used for the orientation factor on the basis of fluorescence polarizations experiments which demonstrated dynamic, restricted movement of both the acceptor and donor relative to the peptide backbone. The parameters and average intramolecular distances in various solvents are given in Table 3. The results show that the average distance is remarkably independent of solvent, and, in particular, there is no dramatic influence of conformation-breaking solvents. Since these findings are indicative of a "random coil" structure for residues 9-21, the end-to-end distance for a random coil dodecapeptide was estimated by the method of Brant and Flory [17]. A value of 29 Å was obtained which is in good agreement with the average distance of 24 Å calculated from the energy transfer experiments. There is also agreement between this value and the lower limit (19 Å) for the distance between Tyr[23] and Trp[9] obtained by Eisinger [18].

VI.   HETEROGENEITY IN DONOR-ACCEPTOR SEPARATION

In the case of conformational heterogeneity of a macromolecule, the donor-acceptor pairs are no longer uniformly separated. In this situation the efficiency of transfer is given by Eq. (18):

$$E = \int_0^\infty f(r) \frac{R_o^6}{R_o^6 + r^6}\, dr \tag{18}$$

where $f(r)$ is the normalized distribution of distances between

TABLE 3

Förster Parameters and Calculated Intramolecular Distances in $[Lys(Dns)^{21}]ACTH_{1-24}$. Taken from Ref. 7

Solvents, conditions[a]	J_{AD} (cm⁶/mole) ($\times 10^{12}$)	κ^2/n^4	φ_D°	E	r [Å]
H_2O	3.02	0.200	0.10	0.45	20.1
Sodium phosphate, 0.1 M, pH 7.0	3.04	0.200	0.09	0.18	24.5
Sodium phosphate, 0.1 M, pH 7.0, 70°C	3.17	0.205	0.05	0.11	24.8
6 M Guanidine · HCl	3.22	0.148	0.09	0.18	23.6
8 M Urea	3.20	0.161	0.17	0.20	26.0
Protected peptide in H_2O	3.02	0.200	0.21	0.64	20.0

[a] The measurements were made at 25°C, unless indicated otherwise.

[b] The protected peptide is $[Lys(Dns)^{21}]ACTH_{1-24}$ with all amino and carboxyl groups protected by Boc- and Bu^t groups, respectively.

donor-acceptor pairs [19]. This situation is of considerable interest since the behavior of certain biopolymers (e.g., linear polypeptide hormones) is best explained by an equilibrium involving many
different conformations [20].

Two approaches have recently been described in the literature
which should permit the determination of f(r) in the situation where
the relative donor-acceptor distances do not change during the lifetime of the donor fluorescence [19,21]. The method of Cantor and
Pechukas [19] aims to obtain f(r) by determining the dependence of
the transfer efficiency on R_o which should be established from a
series of energy transfer experiments. The variation of R_o is proposed to be brought about by using different donor-acceptor pairs,
by changing the solvent, or by modifying the donor quantum yield.
The first method seems to be the most adequate one; however, it
requires the time-consuming preparation of several compounds. The
two other ones seem to be less useful since they will most likely
cause a change in the conformational equilibrium in the course of
the experiment. The method of Grinvald et al. [21] attempts to obtain f(r) from analysis of the donor fluorescence decay which, in
this situation, is no longer governed by a single exponential but
rather described by Eq. [19]:

$$I_D(t) = k \int_0^\infty f(r) \times \exp\{- \frac{t}{\tau_D^o} - \frac{t}{\tau_D^o} \left(\frac{R_o}{r}\right)^6\} \, dr \qquad (19)$$

The use of a distribution function with adjustable parameters is
then suggested in order to obtain information about f(r) on the basis
of Eq. [19] in a closest fit procedure. Haas et al. [22] have recently used this method with success in a study of the distribution
of end-to-end distances in a series of oligopeptides. The results
elegantly demonstrate the feasibility of the technique for studies
in viscous solution and under the condition of a relatively long
donor lifetime (60 ns). In the case of biopolymers in aqueous solution, the successful use of this method may be limited by very short
lifetimes or by a decay of the donor fluorescence in the absence of

the acceptor which is not governed by a single exponential. The
reasons for departure from a single exponential may be found in solvent relaxation, exciplex formation, heterogeneity in the immediate
environment of the fluorophore, presence of a titratable group, or
existence of a small fraction of modified (e.g., oxidized) fluorophores. Both difficulties emerged in preliminary experiments with
the dansylated ACTH derivative, $[Lys(Dns)^{21}]ACTH_{1-24}$. The relative
merit of these methods for the demonstration and quantitation of
conformational heterogeneity of biopolymers in aqueous solution
will have to be established in the future.

ACKNOWLEDGMENT

I thank Dr. Ludwig Brand for the use of his lifetime instrument
and for interesting discussions.

REFERENCES

1. I. Z. Steinberg, in Annual Review of Biochemistry (E. E. Snell,
    ed.), Vol. 40, Ann. Review Inc., Palo Alto, Calif., 1971, pp.
    83-114.

2. R. E. Dale and J. Eisinger, in Biochemical Fluorescence Con-
    cepts (R. F. Chen and H. Edelhoch, eds.), Marcel Dekker, New
    York, 1975, pp.

3. R. H. Conrad and L. Brand, Biochemistry, 7, 777 (1968).

4. R. P. Haugland, J. Yguerabide, and L. Stryer, Proc. Natl. Acad.
    Sci. USA, 63, 23 (1971).

5. K. Beardsley and C. R. Cantor, Proc. Natl. Acad. Sci. USA, 65,
    39 (1970).

6. T. Tao, J. H. Nelson, and C. R. Cantor, Biochemistry, 9, 3514
    (1970).

7. P. W. Schiller, Proc. Natl. Acad. Sci. USA, 69, 975 (1972).

7a. C.-H. Yang and D. Söll, Proc. Natl. Acad. Sci. USA, 71, 2838 (1974).

7b. J. R. Bunting and R. E. Cathou, J. Mol. Biol., 87, 329 (1974).

8. C.-W. Wu and L. Stryer, Proc. Natl. Acad. Sci. USA, 69, 1104 (1972).

9. L. Brand and B. Witholt in Methods of Enzymology (C. H. W. Hirs, ed.), Vol 11, Academic Press, New York, 1967, pp. 776-856.

10. M. Cortijo, I. Z. Steinberg, and S. Shaltiel, J. Biol. Chem., 246, 933 (1971).

11. J. B. Birks, S. Georghiou, and I. H. Munro, J. Phys. B (Proc. Phys. Soc.) Ser. 2, 1, 266 (1968).

12. I. Isenberg and R. D. Dyson, Biophys. J., 9, 1337 (1969).

13. Th. Förster, Ann. Phys., 2, 55 (1948).

14. D. L. Dexter, J. Chem. Phys., 21, 836 (1953).

15. R. Schwyzer and P. W. Schiller, Helv. Chim. Acta, 54, 897 (1971).

16. P. W. Schiller and L. Brand, unpublished experiments.

17. D. A. Brant and P. J. Flory, J. Amer. Chem. Soc., 87, 2788 (1965).

18. J. Eisinger, Biochemistry, 8, 3902 (1969).

19. C. R. Cantor and P. Pechukas, Proc. Natl. Acad. Sci. USA, 68, 2099 (1971).

20. L. C. Craig, J. D. Fisher, and T. P. King, Biochemistry, 4, 311 (1965).

21. A. Grinvald, E. Haas, and I. Z. Steinberg, Proc. Natl. Acad. Sci. USA, 69, 2273 (1972).

22. E. Haas, M. Wilchek, E. Katchalski-Katzir, and I. Z. Steinberg, Proc. Natl. Acad. Sci. USA (1975), in press.

Chapter 6

FLUORIMETRIC KINETIC TECHNIQUES:
CHEMICAL RELAXATION AND STOPPED-FLOW

Thomas M. Jovin
Department of Molecular Biology
Max Planck Institut fuer Biophysikalische Chemie
Goettingen, German Federal Republic

I. INTRODUCTION

A. Scope of Fluorescence Kinetic Methods

The aim of chemical kinetics is the elucidation of reaction
mechanisms. One seeks information about the number and nature of
elementary steps; about the identity and properties of reactants,
intermediates, and products; and about the kinetic and thermodynamic parameters characterizing equilibria and reaction pathways.
Owing to the inherent nature of most kinetic methods, however, in
which processes are measured as a function of time, the tendency is
to concentrate on the determination of the rate parameters alone,
with a consequent loss of potential information.

This chapter describes the current status and future prospects
of instrumentation and analytical methods for kinetic studies by
using the measurement of fluorescence and transmission. The emphasis
is arbitrarily but necessarily on the method of chemical relaxation
as applied in my laboratory, yet the principles are deemed to be
general. The limiting factors in applications related to biological
systems are often the availability and purity of the substances
under investigation. Thus, two major objectives are the optimization
of the overall strategy of experimentation and the extraction of the
maximal amount of information from the data.

The advantages of fluorescence in the study of complex systems
are, primarily, great sensitivity and a large array of parameters
sensitive to different physical properties of molecules and their
environment. For convenience in later discussions, the fluorescence
variables are summarized in Table 1, together with their general
areas of application.

It is evident from Table 1 that the wealth of information potentially available from fluorescence studies may be difficult to
disentangle in the case of complex systems. Kinetic methods are
particularly valuable in such instances because they supply the

additional dimension of time by which overlapping processes may be
resolved. One can consider, in fact, three clocks which define the
experimental resolution: electronic transitions (excited state
lifetimes), diffusion (rotational relaxation times), and chemical
reaction (chemical relaxation times). Thus, fluorescence techniques
find utility in a time range limited only by the lower boundary for
excited state lifetimes of about 0.1 nsec.

The various kinetic methods which employ fluorescence detection
are given in Table 2. In them are subserved practically all the
parameters and applications listed in Table 1. The newest technique
is probably that of fluorescence correlation spectroscopy [17]. The
temporal correlations of concentration fluctuations about equilibrium
are determined by the rates of chemical reaction and diffusion and
thus supply information about both processes. A unique feature of
the method is the requirement for extremely small numbers of molecules ($\sim 10^4$). In the study cited above, the binding of ethidium
bromide to calf thymus DNA was measured. A similar system will be
described subsequently. Another area which has received much impetus
recently owing to technical, analytical, and theoretical innovations
is that of nanosecond fluorescence spectroscopy [7-9]. Its applications range from the study of excited state processes to the determination of macromolecular structure and motion.

As already indicated, the remainder of this chapter will deal
primarily with the method of fluorescence chemical relaxation kinetics. A recent valuable review by Rigler and Ehrenberg [16] deals
with the general application of fluorescence relaxation spectroscopy
to the study of molecular interactions and structure.

II.   THEORY OF CHEMICAL RELAXATION KINETICS:
SYSTEMS OF COUPLED REACTIONS

The technique of chemical relaxation kinetics developed by
Eigen, De Maeyer and co-workers [21] has found increasing application

TABLE 1

Fluorescence Parameters and Processes and Their Areas of Application[a]

Parameter/Process[b,c]	Code	Applications[b,c]	Code
Quantum yield	1	Molecular structure	
		Size, shape, flexibility, motion	A
Spectra		Conformational change	B
Excitation	2	State of aggregation	C
Emission	3		
Polarization: linear [5]	4	Molecular environment	
circular [6]	5	Polarity and viscosity of microenvironment	D
		Accessibility of solute, solvent	E
Kinetic constants and processes		Proximity relationships	F
Excited state lifetime [7,8]	6	Intra-, intermolecular interactions	G
Rotational relaxation [8]	7		
Time-dependent emission spectra [7,9]	8	Binding equilibria	
		Substrate, cofactor, effector binding to	
Resonance energy transfer		enzymes; antibody-hapten and other	
Singlet-singlet [10]	9	ligand-protein interactions	H
Triplet-singlet [11]	10	Dye binding to macromolecules and membranes	I
Triplet-triplet [12,13]	11	Protein-nucleic acid interactions	J
		Solute, solvent binding	K

Dynamics

Chemical reaction, e.g., enzymatic   L
Kinetics of A-K   M
Complex biological transient phenomena,   N
  e.g., nerve conduction

Analytical

Determination of concentration   O
Fluorescence tagging   P

Analysis and separation of cells

Fluorescence flow photometry   Q
Cell separators   R
Microscopic examination of cellular and   S
  subcellular structures

[a]No one-to-one horizontal correlation between parameters and applications is implied. In many applications, rather, several fluorescence variables are used in a complementary manner.

[b]In most instances, both the fluorescence variable and the system under investigation are influenced by perturbations of external factors such as solvents, supporting electrolytes, pH, and the intensive thermodynamic variables (T,P,E).

[c]The following are reviews and monographs on fluorescence topics: general practice and theory [1,112], biological macromolecules and cells [2,111,112], probes [3,112], and a survey of biophysical fluorescence [4].

TABLE 2

Kinetic Techniques Using Fluorescence Detection

Technique	Thermodynamic condition[a]	Time range, sec	Ref.
Photostationary spectroscopy Equilibrium or steady state	1,5		[105]
Slow perturbation of intensive thermodynamic variable or biological state	2,3	$>10^{-3}$	
Mixing and flow (manual, stopped flow, continuous flow)	2-5	$>10^{-4}$	[14,15][c]
Chemical relaxation kinetics	2	$10^{-6}-1$	[16,39][c]
Correlation spectroscopy	1-2	$10^{-5}-1$	[17]
Light perturbation spectroscopy Laser spectroscopy	2-3	$>10^{-12}$ [b]	[18,19]
Flash photolysis	2-3	$>10^{-6}$	[18,20]
Nanosecond spectroscopy Decay of emission	3	$>10^{-10}$	[7,8]
Depolarization kinetics	3	$>10^{-9}$	[8]
Time-dependent emission	3	$>10^{-9}$	[7,9]

[a]The numbers refer to thermodynamic states as follows: equilibrium (1), near equilibrium (2), far from equilibrium (3), presteady state or transient state (4), and steady state (5).

[b]This time resolution refers to excitation and subsequent spectroscopic transitions and obviously not to the fluorescence emission itself.

[c]This chapter.

because of its sensitivity, time resolution, analytical simplicity, and potential for providing detailed information about reaction mechanisms, intermediate species, and the various kinetic and thermodynamic parameters characterizing a given system. A number of excellent monographs, reviews, and papers are available in which the theoretical and experimental aspects of the method are treated [21-31]. Yet, it is appropriate to consider here the principles involved from the standpoint of the objective stated earlier, namely, the maximization of information derived from experiment.

In the case of transient relaxation methods, a chemical system originally at equilibrium is subjected to a sudden perturbation of an intensive thermodynamic variable (generally, temperature in the case of the instrumentation described in this chapter). If the enforced deviations from thermodynamic equilibrium are small, the system approaches a new stable equilibrium under isothermal, isobaric conditions with a temporal response which can be characterized by linearized first-order equations, that is, by discrete exponential decays. The information content of the response is contained, therefore, in the form of a spectrum of relaxation times, a corresponding set of relaxation amplitudes, and the instantaneous signal changes due to the mechanical and spectroscopic consequences of the altered external conditions, i.e., changes in volume and spectral constants. The experimental approach to the resolution of the system usually consists of studying the dependence of the observed parameter (e.g., transmission or fluorescence) upon the concentrations of the reaction participants, optical configuration of the instrument, and external factors such as solvent, solute, pH, and temperature. A successful outcome consists of the concordance of experimentally derived data with the quantitative predictions of one or a limited set of theoretically analyzed and simulated mechanisms. An important corollary which remains unfulfilled only too often is that other potential mechanisms be rigorously excluded.

Most theoretical treatments of chemical relaxation emphasize the development of analytical expressions for the relaxation times

with less regard for the relaxation amplitudes. The latter, however, are sources of significant thermodynamic and physical information such as equilibrium constants, reaction enthalpies, physical
properties of intermediates, and interaction stoichiometries [21-33].
Relaxation rate equations can be couched in terms of concentration
variables [21] or reaction advancements. The latter formalism introduced by Castellan [23] for relaxation times is particularly
convenient and has been extended to relaxation amplitudes by Thusius
[26,31] and by Schimmel [30]. In the following treatment, I present
some new results for the explicit treatment of thermodynamically
dependent reactions, the correlations between concentration and advancement variables, and generalized amplitude expressions.

A. Mechanism

A system of chemical reactions involving N chemical species
(B_i) and R elementary steps can be symbolized by expressions for the
stoichiometries and the rates of reaction:

$$\sum_{i=1}^{N} B_i \cdot \nu_{i\alpha} = 0 \qquad \vec{v}_\alpha = \vec{k}_\alpha \Pi c_m^{-\nu_{m\alpha}} \qquad \overleftarrow{v}_\alpha = \overleftarrow{k}_\alpha \Pi c_n^{\nu_{n\alpha}} \qquad (1)$$

$$\alpha = 1,\ldots, R \qquad m = \text{reactants} \qquad n = \text{products}$$

where $\underline{\nu}$ is the matrix (N x R) of stoichiometric coefficients (taken
as positive for products and negative for reactants), $\vec{v}_\alpha$ and $\overleftarrow{v}_\alpha$ are
the forward and reverse rates for reaction α, respectively, $\vec{k}_\alpha$ and
$\overleftarrow{k}_\alpha$ are the corresponding rate constants, and the c's are molar concentrations. The assumption is made here that solutions are dilute
and that the supporting electrolyte is present in high enough concentration to insure the constancy of activity coefficients. The
apparent equilibrium constant $K_\alpha = \Pi c_i^{-\nu_{i\alpha}}$ in which the index i scans
both reactants and products, or, where noted, the reciprocal quan-
tity, i.e., dissociation constant, is used throughout.

At equilibrium, the chemical species are present at constant concentration $(\bar{c}_i)$ and $\vec{v}_\alpha = \overleftarrow{v}_\alpha = r_\alpha$. We define the diagonal (R) matrix $\underline{r}$ with elements $\delta_{ij} r_i$ [23]. In the following discussion, the inverse and transpose of matrices shown as underscored quantities are indicated by the superscripts -1 and T, respectively, and the order or dimensions are given in parentheses.

B. Rate Equations

The reaction scheme may be regarded as a finite-dimensional vector space of dimension R' equal to the number of linearly independent steps (out of the total of R steps). The essential feature of relaxation kinetic methods is the imposition of perturbations small enough in magnitude such that subsequent reactions rates are proportional to the affinities. The reaction free energy for each step G_α is consequently a function of all the net advancements.

$$\underline{G} = RT/V \cdot \underline{g}\Delta\underline{\xi} = RT \cdot \Delta\ln K \tag{2}$$

where G is the column vector (R) of free energies; $\Delta\underline{\xi} = \underline{\xi} - \bar{\underline{\xi}}$ is the column vector (R) of the net advancements, and $\Delta\ln K$ is the column vector (R) of the enforced perturbations of the apparent equilibrium constants. The symmetric (R) matrix $\underline{g}$ is given by

$$\underline{g} = \underline{v}^T \bar{\underline{c}}^{-1} \underline{v} \tag{3}$$

where $\bar{\underline{c}}^{-1}$ is a diagonal matrix (N) of reciprocal equilibrium concentrations with zero elements for substances whose concentrations do not change under the perturbation, i.e., catalysts or buffered species [23]. The general element of $\underline{g}$ is thus $g_{\alpha\beta} = g_{\beta\alpha} = \Sigma_i v_{i\alpha} v_{i\beta}/\bar{c}_i$ and the reciprocal diagonal elements are equal to the Γ_α factors which denote the "chemical compliance" of the elementary steps equilibrating independently [21,31].

$$\Gamma_\alpha = -RT \left(\delta\xi_\alpha/\delta A_\alpha\right)_{T,P,\xi_{k\neq\alpha}} \tag{4}$$

where A_α is the affinity ($A_\alpha = -G_\alpha$). The units of Γ_α are thus moles of reaction turnover due to a change of RT in the reaction affinity.

In the absence of dependent reactions, the rate equations can be expressed in terms of $R = R'$ concentration variables $\underline{x} = \underline{c} - \underline{\bar{c}}$ or, alternatively, the advancements [23].

$$\underline{\dot{x}} = -\underline{a}\underline{x} \tag{5}$$

$$\Delta\underline{\dot{\xi}} = -\underline{rg}\Delta\underline{\xi} \tag{6}$$

The formulation of Eq. (5) has been preferred historically but has the disadvantage that the coefficient matrix $\underline{a}$ is derived by linearization of rate equations and the explicit introduction of conservation and mass action equilibrium conditions. In Eq. (6), however, the $\underline{r}$ and $\underline{g}$ matrices are obtained by inspection.

In the event that the mechanism contains thermodynamically dependent steps ($R > R'$), i.e., reaction loops, a set of R' independent basis vectors $\underline{x}$ can be chosen and the corresponding rate equations developed as before with the constraints of $R-R'$ additional conservation conditions. In the alternative procedure adopted here, an equivalent basis consisting of R' independent advancements is selected for which the rate equations are

$$\Delta\underline{\dot{\xi}}' = -\underline{r}'(\underline{g}'\Delta\underline{\xi}' + \underline{\hat{g}}\Delta\underline{\hat{\xi}}) = -\underline{r}'\underline{g}'U\Delta\underline{\xi}' = -\underline{b}\Delta\underline{\xi}' \tag{7}$$

The first expression in Eq. (7) can be derived from Eq. (85) of Castellan [23]. The square (R') matrices $\underline{r}'$ and $\underline{g}'$ are the submatrices of $\underline{r}$ and $\underline{g}$, respectively, corresponding to the set of advancements $\Delta\underline{\xi}'$; $\underline{\nu}'$ is the matrix ($N \times R'$) of stoichiometric coefficients; $\underline{\hat{r}}$ is the diagonal matrix ($R-R'$) of dependent equilibration rates; and $\underline{s}$ (with transpose $\underline{s}^T$) is the transformation matrix of dimension $R' \times (R-R')$, relating the dependent to the independent parameters [23]. The following three identities represent Eqs. (29), (27), and (84), respectively of Castellan [23] in matrix form.

$$\underline{\hat{g}} = -\underline{g}'\underline{s} \qquad\qquad R' \times (R-R')$$

$$\underline{\nu} = -\underline{\nu}'\underline{s} \qquad\qquad N \times (R-R') \tag{8}$$

$$\Delta\underline{\hat{\xi}} = -\underline{\hat{r}}\underline{s}^T\underline{r}'^{-1}\Delta\underline{\xi}' \qquad\qquad R-R'$$

The general element of the square (R') transformation matrix $\underline{U}$ which follows from Eqs. (7) and (8)

$$\underline{U} = \underline{I} + \underline{s}\hat{\underline{r}}\underline{s}^T\underline{r}'^{-1} \tag{9}$$

is given by $U_{ij} = \delta_{ij} + r_j^{-1} \sum_{k=1}^{R-R'} s_{ik}s_{jk}\hat{r}_k$. In the absence of dependent reactions, $\underline{U}$ reduces to the identity matrix and Eq. (7) is equivalent to Eq. (6).

The coefficient matrices $\underline{a}$ and $\underline{b}$ are related by the following similarity transformation:

$$\underline{a} = (\underline{\nu}''\underline{U})\underline{b}(\underline{\nu}''\underline{U})^{-1} = \underline{\nu}''(\underline{U}\underline{r}'\underline{g}')\underline{\nu}''^{-1} \tag{10}$$

in which ν'' is the submatrix $(R' \times R')$ of stoichiometric coefficients corresponding to the same set of chemical species comprised in $\underline{a}$.

C. Temporal Response: Relaxation Spectrum

In transient relaxation techniques, such as the temperature jump, the perturbation is regarded most conveniently as the sudden release of a pre-existing displacement from equilibrium since the system parameters under which the relaxation occurs correspond to the final thermodynamic state. Under these conditions, the solutions to Eqs. (5) to (7) assume the form

$$\underline{x}(t) = \underline{A} \cdot \underline{d} \qquad \text{and} \qquad \Delta\underline{\xi}'(t) = \underline{B} \cdot \underline{d} \tag{11}$$

where $\underline{d}$ is the column vector (R') of time-dependent elements

$$d_i = 1 - \lambda_i \exp(-\lambda_i t)*f(t) \qquad (d_i = 1 \text{ at } t = 0) \tag{12}$$

in which the λ_i are the R' reciprocal relaxation times, $f(t)$ is the time-dependent factor of the thermodynamic forcing function, and $\underline{A}$ and $\underline{B}$ are square matrices (R') of coefficients the rows of which sum to yield $\underline{x}_o$ and $\Delta\underline{\xi}_o'$, the column vectors (R') of enforced initial displacements of concentrations and advancements, respectively. The forcing function $f(t)$ appears in the form of a convolution, denoted by the symbol * in Eq. (12), which in the case of the exponential function $f(t) = h(t)[1 - \exp(-\lambda_s t)]$ corresponding, for example, to a condenser discharge, yields

$$d_i = (1 - \lambda_i/\lambda_s)^{-1}[\exp(-\lambda_i t) - (\lambda_i/\lambda_s)\exp(-\lambda_s t)] \qquad (13)$$

As $\lambda_s \to \infty$, $f(t)$ reduces to the step function $h(t)$ and $d_i \to \exp(-\lambda_i t)$. From the standpoint of later discussions of relaxation amplitudes, it is important to note that even though the transient response is relatively insensitive to λ_s for $\lambda_s > \lambda_i$, the pre-exponential term $(1 - \lambda_i/\lambda_s)^{-1}$ may be far from negligible. Thus the relaxation amplitude is more affected than the relaxation time.

The reciprocal relaxation times λ_i are the eigenvalues of the $\underline{a}$ and $\underline{b}$ coefficient matrices and thus are obtained formally by solution of the characteristic equations [21,23].

$$|\underline{a} - \underline{\Lambda}| = 0 \qquad \text{or} \qquad |\underline{b} - \underline{\Lambda}| = 0 \qquad (14)$$

where $\underline{\Lambda}$ is the diagonal eigenvalue matrix (R') and is the same for matrices $\underline{a}$ and $\underline{b}$ by virtue of Eq. (10). The eigenvalues also can be obtained in the course of a normal mode transformation as described below. Since the trace and determinant of similar matrices, as well as the eigenvalues, are invariant under transformation, it follows that

$$\Sigma \lambda_i = \text{tr } \underline{a} = \text{tr } \underline{b} \qquad \text{and} \qquad \Pi \lambda_i = |\underline{a}| = |\underline{b}| \qquad (15)$$

In general, relaxation spectra are analytically discrete but can approach a continuum given a large enough multiplicity of states and species. For some mechanisms, conditions of degeneracy [21,24, 32,58] or vanishing steady-state concentrations of intermediates [23] can reduce drastically the number of observable relaxation processes.

D. Relaxation Amplitudes

The overall relaxation amplitude has been stressed by Thusius [26,31] as an experimentally accessible parameter of great utility.

It is defined as the total signal change (e.g., in transmission or fluorescence) associated with the relaxation spectrum and is thus a scalar quantity

$$\Delta P^O_{tot} = \underline{\varphi}^T \cdot \underline{\Delta c} \tag{16}$$

where $\underline{\varphi}$ is the column vector (N) of "specific signals" ($\varphi_i = \delta P / \delta c_i$) at constant temperature and pressure corresponding to physical parameter P and chemical species i, and $\underline{\Delta c}$ is the column vector (N) of enforced changes in equilibrium concentrations.

$$\underline{\Delta c} = 1/V \cdot \underline{\nu \Delta \xi}_O = 1/V \cdot \underline{\nu' U \Delta \xi}_O' \tag{17}$$

From Eqs. (2) and (8)

$$\underline{\Delta \ln K}' = 1/V \cdot \underline{g' U \Delta \xi}_O' \tag{18}$$

It follows from Eqs. (17) and (18) that

$$\underline{\Delta c} = \underline{\nu' g}'^{-1} \underline{\Delta \ln K}' \tag{19}$$

and thus

$$\Delta P^O_{tot} = 1/V \cdot \underline{\Delta \varphi}^T \underline{U \Delta \xi}_O' = \underline{\Delta \varphi}^T \underline{g}'^{-1} \underline{\Delta \ln K}' \tag{20}$$

where $\underline{\Delta \varphi}$ is the column vector (R') of specific signal changes corresponding to the R' elementary steps [31]. Thus $\underline{\Delta \varphi} = \underline{\nu}'^T \underline{\varphi}$.

Equation (20) indicates that ΔP^O_{tot} is not a function of kinetic parameters, i.e., rate constants, nor is it affected by the introduction of dependent reactions. The overall amplitude measured as a function of concentration(s) is useful for the experimental determination of the thermodynamic functions $\underline{\Delta \ln K}'$, assuming that the $\Delta \varphi_i$'s and K's for the system are known [26,31].

In general, the relaxation spectrum consists of a sum of R' exponential decays given by

$$\underline{\Delta P}(t) = \underline{d} \cdot \underline{\Delta P}^O \tag{21}$$

where $\underline{d}$ is a diagonal matrix (R') with elements $\delta_{ij} d_i$ [Eq. (12)] and $\underline{\Delta P}^O$ is the column vector (R') of individual relaxation amplitudes.

The relaxation processes are kinetically uncoupled if the corresponding relaxation times are widely enough separated on the time axis. The analysis of relaxation times and amplitudes is greatly simplified under this circumstance as will be discussed later.

Relaxation times that are of the same order of magnitude result from kinetic coupling between reactions. In such instances, recourse can be made to a "normal mode" transformation. Either of the coefficient matrices $\underline{a}$ or $\underline{b}$ of Eqs. (5) and (7) can be diagonalized by a similarity transformation $\underline{Q}^{-1}\underline{a}\underline{Q}$ or $\underline{Q}^{-1}\underline{b}\underline{Q}$ where $\underline{Q}$ ($R' \times R'$) is constructed from the normalized eigenvectors of $\underline{a}$ or $\underline{b}$, respectively. The bases $\underline{x}$ or $\underline{\Delta\xi}'$ are thereby transformed to a new set of totally uncoupled vectors. Excellent discussions of this general problem are given by Eigen and De Maeyer [21] and Hayman [22]. The following procedure represents a new attempt to optimize the choice of normal coordinates in the general case of kinetically (and necessarily thermodynamically) coupled sets of reactions, some of which in addition may be thermodynamically dependent.

In the first step, the independent $\underline{\Delta\xi}'$ are transformed by $\underline{U}$ [Eq. (9)] to a new set of advancements $\underline{\Delta\xi}^O$ and corresponding rate equations.

$$\underline{\Delta\xi}^O = \underline{U}\underline{\Delta\xi}' \tag{22}$$

$$\underline{\dot{\Delta\xi}}^O = -(\underline{U}\underline{r}')\underline{g}'\underline{\Delta\xi}^O \tag{23}$$

The product $\underline{U}\underline{r}'$ is symmetric and thus can be diagonalized by an orthogonal transformation.

$$\underline{M}^T(\underline{U}\underline{r}')\underline{M} = \underline{D} \qquad (\underline{M}^T\underline{M} = \underline{I}) \tag{24}$$

The matrix $\underline{M}^T$ is used to transform the $\underline{\Delta\xi}^O$ to a set of vectors $\underline{\Delta z}$, also with the dimensionality of advancements (moles).

$$\underline{\Delta z} = \underline{M}^T\underline{\Delta\xi}^O \tag{25}$$

Thus

$$\underline{\dot{\Delta z}} = -\underline{D}\underline{J}\underline{\Delta z} \tag{26}$$

where

$$\underline{J} = \underline{M}^T \underline{g}' \underline{M} \tag{27}$$

with order R'. Since $\underline{D}$ is diagonal and $\underline{J}$ symmetric, a further and final orthogonal transformation to the normal coordinates $\underline{\Delta y}$ can be performed in the manner proposed by Castellan for the $\underline{rg}$ matrix product [23]. Thus

$$\underline{\Delta y} = \underline{W}^T \underline{D}^{-1/2} \underline{\Delta z} \qquad [\text{units} = (\text{moles-liter-sec})^{1/2}] \tag{28}$$

and

$$\underline{\dot{\Delta y}} = -[\underline{W}^T (\underline{D}^{1/2} \underline{J} \underline{D}^{1/2})\underline{W}]\underline{\Delta y} = -\underline{\Lambda}\ \underline{\Delta y} \tag{29}$$

where $\underline{W}$ is the orthogonal matrix (R') which diagonalizes the symmetric product $\underline{D}^{1/2} \underline{J} \underline{D}^{1/2}$, thus leading to $\underline{\Lambda}$, the matrix of reciprocal relaxation times [Eqs. (14) and (29)].

The square transformation (R') matrix $\underline{T}$ relates $\underline{\Delta \xi}^o$ to $\underline{\Delta y}$.

$$\underline{\Delta \xi}^o = \underline{T} \underline{\Delta y} \tag{30}$$

$$\underline{T} = \underline{M} \underline{D}^{1/2} \underline{W} \qquad \text{and} \qquad \underline{T}^T = \underline{W}^T \underline{D}^{1/2} \underline{M}^T \tag{31}$$

In order to obtain $\underline{\Delta \varphi}^y$, the column vector (R') of specific signal changes corresponding to the normal coordinates $\underline{\Delta y}$, one notes that

$$\Delta P_{tot}^o = 1/V \cdot (\underline{\Delta \varphi}^y)^T \cdot \underline{\Delta y}_o = 1/V \cdot \underline{\Delta \varphi}^T \underline{\Delta \xi}_o^o \tag{32}$$

where the first expression represents the sum of individual amplitudes, $1/V \cdot (\Delta \varphi^y)_i (\Delta y_o)_i$, and the second expression arises from Eqs. (20) and (22). From Eqs. (30) and (32)

$$\underline{\Delta \varphi}^y = \underline{T}^T \underline{\Delta \varphi} = \underline{T}^T \underline{v}^T \underline{\varphi} \tag{33}$$

In seeking an expression for the enforced displacements $\underline{\Delta y}_o$, one uses the fact that the overall free energy change for the system is given by

$$\Delta G = -\underline{A}_o^T \underline{\Delta \xi}_o^o = -\underline{Y}_o^T \underline{\Delta y}_o \tag{34}$$

where Y is the cognate (to y) thermodynamic variable analogous to
the affinity A in its relation to the advancement ξ. From Eqs. (30)
and (34)

$$\underline{Y}_o = \underline{T}^T \underline{A}_o \tag{35}$$

and from Eqs. (18) and (22)

$$\underline{A}_o = -RT/V \cdot \underline{g}' \Delta \underline{\xi}_o^o = - RT \cdot \underline{\Delta \ln K}' \tag{36}$$

We define the column vector $\underline{\Gamma}^y$ of normal Γ factors (R')

$$\Gamma_j^y = -RT(\partial y_j/\partial Y_j)_{T,P,y_{k \neq j}} \tag{37}$$

and thus

$$\underline{\Delta y}_o = -1/RT \cdot \underline{\Gamma}^y \underline{Y}_o \tag{38}$$

By substituting Eqs. (35), (36), and (30) into Eq. (38), then using
Eqs. (31), (27), and (29), it can be shown that

$$\underline{\Gamma}^y/V = \underline{\tau} = \underline{\Lambda}^{-1} \tag{39}$$

and thus

$$\underline{\Delta y}_o = V \cdot \underline{\tau} \underline{T}^T \underline{\Delta \ln K}' \tag{40}$$

From Eqs. (32) and (40) we obtain the expression for the relaxation amplitudes in final form

$$\underline{\Delta P}^o = \underline{\tau} \underline{\Delta \varphi}^y \underline{T}^T \underline{\Delta \ln K}' \tag{41}$$

where $\underline{\Delta \varphi}^y$ is a diagonal matrix (R') with the general element $\delta_{ij} \Delta \varphi_i^y$
[Eq. (33)]. Thus the individual products of relaxation times and
amplitudes assume the simple form

$$\underline{\Lambda} \underline{\Delta P}^o = \underline{\Delta \varphi}^y \underline{T}^T \underline{\Delta \ln K}' \tag{42}$$

or

$$\Lambda_i \cdot \Delta P_i^o = \sum_{j=1}^{R'} T_{ij}^T \Delta \varphi_j \sum_{j=1}^{R'} T_{ij}^T \Delta \ln K_j' \tag{43}$$

The above formulation offers many advantages in that (a) it accounts for any degree of kinetic and thermodynamic coupling; (b) the normal Γ factors are related simply to the relaxation times and thus need not be calculated independently. That such is the case for transformations of the form given in Eq. (28) was pointed out by Hayman in 1973 [22]; (c) once having calculated the transformation matrix $\underline{T}^T$, the amplitude expressions can be written by inspection and the $\Delta\varphi$ and $\Delta\ln K$ terms appear symmetrically [Eq. (43)]; and (d) the factors in Eq. (43) can be expanded into sums appropriate for a nonlinear regression analysis of experimental data in a manner similar to that proposed by Thusius [26,31].

In the general case, the various matrices that are required to evaluate Eq. (43) are computed in the order $\underline{M}$ and $\underline{D}$ [Eq. (24)], $\underline{J}$ [Eq. (27)], $\underline{W}$ and $\underline{\Lambda}$ or $\underline{\tau}$ [Eq. (29)], $\underline{T}^T$ [Eq. (31)], and $\underline{\Delta\varphi}^y$ [Eq. (33)].

In the absence of thermodynamically dependent reactions, $\underline{U} = \underline{I}$, $\underline{D} = \underline{r}$, $\underline{M} = \underline{I}$, $\underline{J} = \underline{g}$, $\underline{T}^T = \underline{W}^T\underline{r}^{1/2}$, and thus only $\underline{W}$ and $\underline{\Lambda}$ or $\underline{\tau}$ are required. If, in addition, the relaxation processes are widely separated on the time axis, that is, are kinetically uncoupled, further simplifications are possible. If one numbers the reactions such that the diagonal terms of the $\underline{rg}$ matrix are in descending order, i.e., reaction 1 is the fastest, the relaxation times are given by

$$\lambda_\alpha = \tau_\alpha^{-1} = r_\alpha |\underline{g}_{\alpha,\alpha}| / |\underline{g}_{\alpha-1,\alpha-1}| \qquad (|\underline{g}_{0,0}| = 1) \qquad (44)$$

where $\underline{g}_{k,k}$ is the kth principal submatrix of $\underline{g}$ [23]. It is seen from Eq. (44) that for well-separated relaxation times, the value of λ_α depends upon the kinetic constants of the αth reaction but only the thermodynamic parameters of the faster $\alpha-1$ reactions [23]. Under the above assumptions, only the affinity of the αth reaction is finite during the αth relaxation. The faster reactions equilibrate instantaneously and their equilibrium advancements shift accordingly, while the R-α slower steps in the mechanism are not affected at all.

From Eq. (34)

$$\underline{g}_{\alpha-1,\alpha} \cdot \underline{\Delta\xi}_\alpha = 0 \tag{45}$$

where $\underline{\Delta\xi}_\alpha$ is the column vector (α) of the enforced changes in advancements during the αth relaxation. We define the column vector $\underline{Q}_\alpha$ with elements $Q_{\alpha j} = \Delta\xi_{\alpha j}/\Delta\xi_{\alpha\alpha}$ $(j < \alpha)$, $Q_{\alpha\alpha} = 1$, and thus

$$\underline{g}_{\alpha-1,\alpha-1} \cdot \underline{Q}_\alpha = -\underline{g}_\alpha \tag{46}$$

where $\underline{g}_\alpha$ is the αth column vector of $\underline{g}_{\alpha-1,\alpha}$ and only the first $\alpha-1$ elements of $\underline{Q}_\alpha$ are used. It follows that

$$\underline{Q}_\alpha^T = -\underline{g}_\alpha^T \underline{g}_{\alpha-1,\alpha-1}^{-1} \qquad (Q_{\alpha\alpha}^T = 1) \tag{47}$$

For numerical computation, other methods for the solution of Eq. (45) are available which do not require the inverse matrix in Eq. (47) explicitly. $\underline{Q}_\alpha$ is the transformation vector corresponding to the selection of $\Delta\xi_{\alpha\alpha}$ as the αth normal advancement. Thus from Eqs. (28) and (30), $\underline{T}_\alpha \Delta y_\alpha = \underline{Q}_\alpha \Delta\xi_{\alpha\alpha}$, $\Delta y_\alpha = r_\alpha^{-1/2}\Delta\xi_{\alpha\alpha}$, and thus

$$\underline{T}_\alpha^T = r_\alpha^{1/2}\underline{Q}_\alpha^T \tag{48}$$

The final expression for the relaxation amplitudes is derived from Eqs. (43), (44), and (48).

$$\Delta P_\alpha = \frac{|\underline{g}_{\alpha-1,\alpha-1}|}{|\underline{g}_{\alpha,\alpha}|} \sum_{j=1}^\alpha Q_{\alpha j}^T \Delta\varphi_j \sum_{j=1}^\alpha Q_{\alpha j}^T \Delta\ln K_j \tag{49}$$

The relaxation amplitudes for a dispersed spectrum are therefore independent of kinetic factors, although Eqs. (42) and (43) are still valid with T_{ij}^T and τ_i being given by Eqs. (48) and (44), respectively.

The following interesting and simple relationships follow from Eqs. (44), (48), and (49).

$$\Gamma_\alpha^\xi/V = |\underline{g}_{\alpha-1,\alpha-1}|/|\underline{g}_{\alpha,\alpha}| = \tau_\alpha r_\alpha \tag{50}$$

where Γ_α^ξ corresponds to the αth normal advancement [Eq. (4)]. The second identity in Eq. (50) is particularly useful in the computational and analytical use of Eq. (49). It can be shown further that the complete ($R' \times R'$) transformation matrix $\underline{Q}$ is of unit upper triangular form and fulfills the relationship given by Schimmel [30] $\underline{Q}^T\underline{g}\underline{Q} = (\underline{\Gamma}^\xi/V)^{-1}$ in which $\underline{\Gamma}^\xi$ is the diagonal matrix of normal Γ factors. The more general transformation matrix $\underline{T}$ [Eq. (31)] operating upon $\underline{g}'$, i.e., $\underline{T}^T\underline{g}'\underline{T}$, leads to the same result.

The expressions for the relaxation times and amplitudes given in Eqs. (44) and (49) can be evaluated by inspection for many mechanisms and are therefore useful for the derivation of analytical relationships.

It remains to add that for the temperature jump,

$$\Delta \ln K = -\Delta H/RT^2 \cdot \Delta T \qquad\qquad (51)$$

where ΔH is the reaction enthalpy.

E.   Total Signals and
Instantaneous Signal Changes

The rapid perturbation of a chemical system results in signal changes secondary to readjustments of physical parameters. Such effects are unrelated to the chemical relaxation processes which ensue and yet can contribute a significant if not predominant component to the overall signal change. In the case of the temperature jump, the thermal expansion or contraction shall be considered to be coincident with the heating. Thus, the rapid initial signal component reflects the changes in volume and the temperature dependencies of the physical parameters being measured.

The total signal S before the perturbation can be represented in a general form applicable to transmission or fluorescence measurements:

$$S = AI_o \cdot F(P,P_o) \tag{52}$$

where A is an overall machine constant, I_o is the incident light intensity, and F is a function, different for transmission and fluorescence, in which P and P_o are the intrinsic chemical and incidental signal sources, respectively. In fluorescence, assuming a low absorbance and thus a signal proportional to concentration,

$$F = bP + b_o P_o$$

$$P = \sum \varphi_i \bar{c} = L\Gamma_{\parallel} + (1 - L)\Gamma_{\perp}$$

$$\Gamma_{\parallel} = \sum \epsilon_i q_i \gamma_{\parallel} \ \bar{c}_i \tag{53}$$

$$\Gamma_{\perp} = \sum \epsilon_i q_i \gamma_{\perp} \ \bar{c}_i$$

$$\varphi_i = \epsilon_i q_i [L\gamma_{\parallel} + (1 - L)\gamma_{\perp}]$$

In the above expressions, b is a machine constant including the terms defining the optical transfer efficiency for the excitation and emission; P_o contains all background sources of light, such as solvent fluorescence, and scattered light with the corresponding coefficient b_o; ϵ_i and q_i are the molar extinction coefficient and quantum yield of species i, respectively; $\gamma_{\parallel}$ and $\gamma_{\perp}$ are, for species i, the ratios of (a) the intensities of emission along axes parallel and perpendicular, respectively, to the direction of linearly polarized excitation, to (b) the sum of the intensities along the three rectangular coordinates ($\gamma_{\parallel} + 2\gamma_{\perp} = 1$, by symmetry); and L is a machine constant, described further below, which includes terms for the polarization of emission and/or excitation and the optical aperture of observation.

In absorbance, with S being a measure of the transmitted light,

$$F = \exp[\ell(P + P_o)] + P' \tag{54}$$

$$P = \sum \varphi_i \bar{c}_i = \sum \epsilon_i \bar{c}_i$$

where ℓ is the light path, P_o represents background absorption, and P' is the stray light contribution, assumed to be constant.

The instantaneous signal change upon heating, ΔS_{in}, is given for small perturbations by

$$\Delta S_{in} = AI_o[\partial F/\partial P \cdot \Sigma \, \partial P/\partial h_i \cdot dh_i/dT$$

$$+ \partial F/\partial P' \cdot dP'/dT]\Delta T \tag{55}$$

where h_i are the various temperature-dependent physical parameters in P.

For fluorescence measurements,

$$\Delta S_{in} = \left\{ (S - S_o)\left[\frac{\partial \ln\rho}{\partial T} + \Sigma \, \varphi_i \bar{c}_i \, \frac{\partial \ln\varphi_i}{\partial T} \bigg/ \Sigma \, \varphi_i \bar{c}_i \right] \right.$$

$$\left. + \frac{\partial \ln S_o}{\partial T} \right\} \Delta T \tag{56}$$

where

$$\frac{\partial \ln\varphi_i}{\partial T} = \frac{\partial \ln\epsilon_i}{\partial T} + \frac{\partial \ln q_i}{\partial T} + \frac{\partial}{\partial T} \, [L\gamma_\| + (1 - L)\gamma_\perp]$$

ρ is the solvent density, and S_o is the background signal due to P_o [Eq. (53)]. The first term in Eq. (56) represents the volume change, the second specifies the contribution from changed spectral constants and usually predominates owing to the characteristically large temperature dependence of fluorescence quantum yields, and the third term provides for possible shifts in background emission.

For transmission measurements,

$$\Delta S_{in} = - \left\{ (S - S_\infty)E\left[\frac{\partial \ln\rho}{\partial T} + \Sigma \, \epsilon_i \bar{c}_i \, \frac{\partial \ln\epsilon_i}{\partial T} \bigg/ \Sigma \epsilon_i \bar{c}_i \right] \right.$$

$$\left. + \frac{\partial E_o}{\partial T} \right\} \Delta T \tag{57}$$

where S_∞ is the stray light signal, measured in the presence of infinite absorbance at the given wavelength, and E and E_o are the absorbance (to the base e) of the composite chemical system and that of residual components such as solutes, respectively.

Fast signal changes preceding the observed relaxation process
may also occur owing to the rapid equilibration of protolytic reactions. Thus, in a buffered solution, a pH change may accompany the
heating and alter the chemical composition of the system to an arbitrary degree. Such effects have to be treated formally as relaxation processes.

It follows from Eqs. (52) to (57) that the corrected total
signal, S^O, is given by

$$S^O = S - S_o \qquad\qquad \text{fluorescence}$$

$$S^O = (S - S_\infty) \exp(E_o) \qquad \text{transmission}$$

(58)

III. KINETIC SPECTROFLUORIMETER

In the development of the instrumentation described in this
section, the objectives were to provide for the following techniques
using constant illumination: (a) steady state titration, (b) temperature-jump relaxation kinetics, (c) stopped-flow kinetics, (d)
steady state and transient kinetics, (e) combinations of (c-d)
with (b), and (f) use of either transmission or fluorescence parameters with flexible optical configurations. Additional features
which are being incorporated currently are the measurement of light
scattering [32], application of a tunable dye laser as a second
light source for flash photolysis [36], the development of a cell
for pressure-jump experiments [37], and the extensive use of computer control and automated data acquisition and analysis. A number
of individuals have been involved in this work, and a technical description of the basic temperature-jump apparatus has been published
elsewhere [39]. It is also necessary to note previous reports of
instrumentation for fluorescence relaxation [40] and flow [41-54]
kinetic measurements.

A. Spectroscopic
Parameters and Optical Configuration

The optical coordinate system of the apparatus is shown in Fig. 1. Four photomultipliers, A, B, C, D, are located in such a way as to enable the recording of signals $\underline{A}$, $\underline{B}$, $\underline{C}$, $\underline{D}$ proportional to the quantities: $(\underline{C})$:I_o (excitation intensity); $(\underline{D})$:I (transmitted intensity); and $(\underline{A}, \underline{B})$:$F_A$, F_B (intensities of fluorescence measured by two photomultipliers A and B. Cutoff or interference filters determine the spectral bandwidth of light impinging on the photocathodes. In the presence of emission analyzers, F_A (θ), F_B (ψ) is the fluorescence detected by A and B through polarizers set at angles θ and ψ, respectively. These signals can be combined electronically and processed in ways described below.

Signals $\underline{D}$ and $\underline{A}$ (or $\underline{B}$) are given by Eqs. (53) and (54), and it follows that the responses $\underline{\Delta D}$, $\underline{\Delta A}$ $(\underline{\Delta B})$ to perturbations of P, the function of chemical and molecular parameters, are

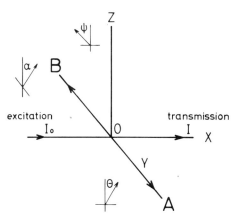

FIG. 1. Optical coordinate system. The sample cell is situated at the origin O. Excitation is with monochromatic light with the option of an additional polarizer or depolarizer. Fluorescence emission is observed at right angles to the excitation, i.e., along the Y axis. Two detectors (A, B) are situated on either side of the cell. The angular settings of an optical excitation polarizer (α) or emission analyzers (θ, ψ) are measured relative to the Z axis.

$$\ln[1 + \underline{\Delta D}/\underline{D}^{O}] = -\ell\Delta P_{T} \cong \underline{\Delta D}/\underline{D}^{O} \tag{59}$$

$$\underline{\Delta A} = AI_{o}b_{A}\Delta P_{F} \quad [\text{Eqs. (52), (53)}] \tag{60}$$

where

$$\Delta P_{T} = \Sigma\epsilon_{i}\Delta c_{i} \quad [\text{Eq. (54)}] \quad \text{transmission} \tag{61}$$

$$\Delta P_{F} = L_{A}\Delta\Gamma_{\parallel} + (1 - L_{A})\Delta\Gamma_{\perp} \quad [\text{Eq. (53)}] \quad \text{fluorescence} \tag{62}$$

and $\underline{D}^{O}$ is given by Eq. (58). The right-hand expression in Eq. (59) is the linear approximation usually fulfilled in transmission re-laxation experiments but not in general. The sensitivity of fluor-escence measurements derives from the direct proportionality between chemical and signal changes, Eq. (60), whereas in transmission the corresponding relationship involves the relative signal.

The significance of Eqs. (60) and (62) is made more evident in Table 3. By suitable selection of the optical configuration, the **fluorescence signals can be made to reflect the following:**

1. Total Fluorescence Emission

This condition is achieved whenever $P_{F} \propto (\Gamma_{\parallel} + 2\Gamma_{\perp})$, because from Eq. (53) φ_{i} reduces to $\epsilon_{i}q_{i}$ and thus $P_{F} = \Sigma\epsilon_{i}q_{i}c_{i}$. The signal is thus independent of the polarizarion of emission. Of the various possibilities indicated in Table 3, probably the most useful is the case of an excitation polarizer set to the indicated angle for $s = 2/3$. In this instance, emission analyzers are not required, and the signal is additionally insensitive (except in magnitude) to the optical aperture of observation. In the absence of any polarizer, the response function is undefined, since the linear polarization of light exiting from a monochromator is highly variable. Assuming depolarized excitation and no emission analyzer, $P_{F} \propto (\Gamma_{\parallel} + 3\Gamma_{\perp})$ (Table 3) and thus includes molecular orientation factors.

2. Emission Polarization or Anisotropy

The three parameters commonly used to express the asymmetry of fluorescence emission are:

TABLE 3

Dependence of Fluorescence Response Function P_F[a]

on the Optical Configuration of Excitation and Emission

	Excitation parameter s[b]			
	0	1/2[c]	2/3	1
Emission aperture[d]	(1 - L)/L[a] (no emission analyzer)[e]			
Small	∞	3	2	1
Large	8.9	2.64	2	1.23

L	P_F	Orientation angle of emission analyzer[f]			
1	$\Gamma_\parallel$	-	-	-	0(*)
1/2	$\frac{1}{2}(\Gamma_\parallel + \Gamma_\perp)$	-	0(*)	30(27.7)	45(41.7)
1/3	$\frac{1}{3}(\Gamma_\parallel + 2\Gamma_\perp)$	-	35.3(37.2)	45(45)	54.7(52.4)
0	$\Gamma_\perp$	0-90(*)	90(*)	90(*)	90(90)

[a] P_F is defined as $L\Gamma_\parallel + (1 - L)\Gamma_\perp$, Eq. (53).

[b] s is the ratio of the vertically polarized intensity to the total intensity produced by an ideal excitation polarizer set to angle α (Fig. 1). For s = 0, 1/2, 2/3, and 1, α = 90, 45, 35.3, and 0 deg, respectively.

[c] This condition is also achieved by the use of a depolarizer.

[d] A small aperture refers to the limit as the solid angle of observation tends to 0. A large aperture refers to a cone of observation with a half-angle of 34 deg; the values are theoretically calculated.

[e] Under these conditions, the factor b in Eq. (53) includes as a multiplicative factor the transmission of the polarizer (or depolarizer) used in excitation.

[f] The indicated angles apply to small apertures (in parenthesis to large apertures). A star indicates that the condition is not feasible. The factor b in Eq. (53) includes the transmission of the emission analyzer (which is assumed to have an infinite contrast ratio) in addition to the quantities referred to in footnote e.

Polarization $\qquad$ $p = (\Gamma_\parallel - \Gamma_\perp)/(\Gamma_\parallel + \Gamma_\perp)$

Emission anisotropy $\qquad$ $A = (\Gamma_\parallel - \Gamma_\perp)/(\Gamma_\parallel + 2\Gamma_\perp)$ $\qquad$ (63)

Degree of polarization $\quad r = \Gamma_\perp/\Gamma_\parallel)$

From Table 3, it can be seen that the polarization parameters can be formed from suitable combinations of the signals $\underline{A}$ and $\underline{B}$. One common configuration is the use of vertically polarized excitation (S = 1) and emission analyzers such that $\theta = 0$ and $\psi = 90$ deg (Fig. 1). The ratios $(\underline{A} - \underline{B})/(\underline{A} + \underline{B})$ or $(\underline{A} - \underline{B})/(\underline{A} + 2\underline{B})$ are then obtained electronically. The apparent parameters p' and A' so derived, however, have to be corrected for the effect of the finite emission aperture. Other combinations, however, permit the simultaneous determination of the true polarization parameters and the total emission as in (1) above. In one useful method, an excitation polarizer is set for s = 2/3, and an emission analyzer is placed in the path of photomultiplier A. Signals $\underline{A}$ and $\underline{B}$ are then matched electronically with $\theta = 45$ deg (Table 3). Angle θ is then set to 27.7 deg (Table 3 for large emission apertures) so that

$\underline{A} \propto (\Gamma_\parallel + \Gamma_\perp)/2$

$\underline{B} \propto (\Gamma_\parallel + 2\Gamma_\perp)/3 \quad$ total emission

$(\underline{A} - \underline{B})/\underline{A} = p/3 \quad$ polarization $\qquad\qquad$ (64)

$(\underline{A} - \underline{B})/\overline{\underline{B}} = A/2 \quad$ anisotropy

The above formalism provides a method for correlating signals and the desired quantities ΔP_T and ΔP_F [Eqs. (61), (62)]. Their use in titration and conventional kinetics requires no elaboration. In relaxation experiments, however, ΔP_T and ΔP_F correspond to the relaxation amplitudes defined by Eqs. (20), (41) and are related through Eqs. (59) and (60) to the measured signal changes $\underline{\Delta A}$, $\underline{\Delta B}$, or $\underline{\Delta D}$ (or more generally, $\underline{\Delta S}$). The experimental relaxation amplitude, M, is henceforth understood to consist of the quantity ΔP_T in the case of transmission and ΔS directly in the case of fluorescence. In the latter instance, however, the signals must be scaled so as to conform to a standard reference state with respect to the factors A, I_o, and b in Eq. (60).

A system of reactions will produce fluorescence signals if
there are changes in any of the parameters or processes indicated
in Table 1. In the case of energy transfer, specifically, the
structure containing each donor and acceptor pair can be formally
treated as one species, with the particular combination of excitation and emission spectra reflecting the presence of quenching and/or
sensitized emission. The processes of concentration quenching and
depolarization are often susceptible to this form of interpretation
as well, thereby simplifying the treatment of spectral coefficients
which might otherwise be concentration dependent. Thus, in all
cases, any linear combination of $\Gamma_{\parallel}$ and $\Gamma_{\perp}$ (Table 3) fulfills the
criterion [Eq. (20)] that the optical property must be a linear
function of concentration. The polarization parameters, however,
are rational polynomials. Their direct use requires either a complete knowledge of the system or the condition of small perturbation such that linearization of Eq. (63) is possible. Under such
circumstances, it can be shown that changes in either p, A, or r
are proportional to a particular linear combination of $\Delta\Gamma_{\parallel}$ and $\Delta\Gamma_{\perp}$,
namely $r\Delta\Gamma_{\parallel} - \Delta\Gamma_{\perp}$. Thus, one requires a signal proportional to
$r\Gamma_{\parallel} - \Gamma_{\perp}$ which, by virtue of Eq. (63) must be 0 before the perturbation. By superposition, any set of signals $\underline{A}$ and $\underline{B}$ in which $\Gamma_{\parallel}$
and $\Gamma_{\perp}$ are weighted differently and for which $a\underline{A} + b\underline{B} = 0$ will qual-
ify. Ratios [Eq. (64)] may be used but are thus not required for
the dynamic measurement, a fact of some experimental importance.

B. Apparatus

The kinetic instrument for fluorescence and transmission measurements consists of four parts: optical system, high-voltage discharge circuit, drive unit for flow measurements, and associated
electronics including analog signal conditioning and computer control of operation and data acquisition.

The optical system is shown schematically in Fig. 2. The whole
assembly is mounted on optical rails with rigidly defined optical

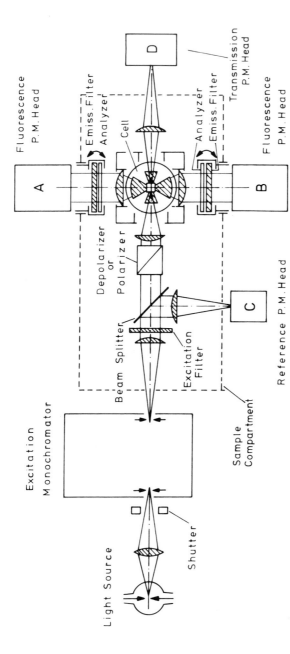

FIG. 2. Optical system of kinetic apparatus. The light sources are generally high-intensity arc or filament lamps, and for special applications, continuous lasers are used. The sample compartment contains a coaxial connector through which the high-voltage source is made available to the cell. All lenses and the beam splitter are constructed from strain-free Suprasil or Spectrosil. The polarizing elements are the following: Hanle-type depolarizer (Halle, Berlin) with transmission better than 83% at 254 nm; Glan-Thompson polarizer (Halle, Berlin) with a transmission for unpolarized light of 11, 20, and 42% at 254, 280, and 365 nm, relatively; analyzers are the uv sheet films Ks-W 68 (Kaesemann, Oberaudorf) with a transmission and contrast ratio of 20% and 10^4 at 350 nm and 40% and 10^5 at 500 nm. The photomultipliers are in intimate contact with the sample compartment. Usually A and B are the EMI 9558-Q (2-in. end-on, S-20 cathode) and C and D the RCA 1P28 or EMI 9781. There is also provision for front-face observation of fluorescence.

axes. The cell fits into a thermostated compartment of standardized dimensions. A number of different designs allows for temperature-jump, stopped-flow, and combined stopped-flow and temperature-jump measurements. Two examples of cells are shown in Figs. 3 and 4. Similar microcells are also available. The common features and objectives in all cases have been high aperture of collection for the fluorescence light, easy access to the cell interior for ease in titration experiments, minimal volume for conservation of material, freedom from materials potentially harmful to biological systems, and strain-free optics and mounting for accurate polarization measurements. The functional characteristics of some representative cells are given in Table 4. In addition, virtually all former cells constructed for transmission measurements [21] are also compatible with the instrument. Standard cuvettes can be accommodated for titration experiments or for purposes of standardization by use of an adapter which fits in the cell compartment.

The high-voltage discharge system is of standard design. New features are the interchangeability of condensers and the optional use of a shunt cell to reduce the discharge time constant for better time resolution.

The drive unit for flow experiments consists of a modified commercial unit (Aminco-Morrow stopped-flow apparatus, American Instrument Co., Silver Spring, Maryland) in which a pneumatic cylinder is actuated by a computer-controlled solenoid. The assembly is mechanically isolated from the optical system, and the drive and stop syringes connect to the flow cells (Fig. 4) through flexible Teflon tubing. The dead time is less than 2.5 msec.

The electronic system is shown in Fig. 5. The signal conditioner, which is extensively described elsewhere [39], can perform the analog functions of subtraction, addition, multiplication, division, and filtering in various modes: low-pass and delayed or undelayed high-pass. The latter provision enables the suppression of rapid transients such as occasional light flashes from the electrodes, the

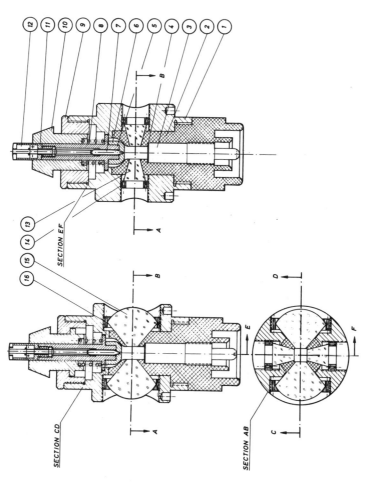

FIG. 3. Standard ("Cyclops") temperature-jump cell. The outer steel shell (1) encloses a plastic body (2). In the center is the cavity or cell proper which is bounded on the bottom by a platinum disk (4) attached to a lower electrode (3) and on the top by an upper removable electrode (6) likewise clad with platinum or gold. The electrode is held in place by a spring-loaded (8) bayonet assembly (10) which also contains a thermistor (7) to monitor the cell temperature. The optical paths through the cell (Section AB) are through plane-faced conical windows (Section AB) in the direction of excitation and convex-conical windows in the direction of emission; the latter have an angular aperture of 80%

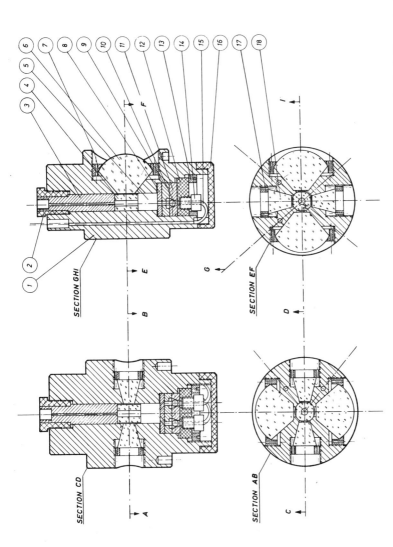

FIG. 4. Stopped-flow cell. The two solutions which are to be mixed are conducted through the cell through Teflon tubing (15) with enough volume for at least one flow-through, thereby constituting a reservoir to insure temperature equilibration. The outer body is metal and the measuring cavity a quartz cylinder (4) with an inner radius of 2.5 mm and outer surfaces ground planar. The solutions enter a multijet Gibson [59] mixer (5) after passing through ball valves (10) which impede back-flow. The mixed solution exits the cell through an assembly (3) connected to Teflon tubing. The remaining optical components are the same as those described in Fig. 3. This cell was designed by H. Lehrach.

TABLE 4

Characteristics of Two Temperature-Jump Cells

Parameter	Cell	
	Cyclops[a]	Microcell[b]
Excitation path length, mm	7-square	3.83 (2.5)-round
Heated volume, ml[c]	0.7	0.17 (0.075)
ΔT, aq. solution, deg/J[d]	0.33	1.35 (-)
Resistivity cell constant[e]	2.7	11.2 (-)
Cooling rate, %/sec[f]	0.3	0.3
Polarization contrast ratio[g]		
Convex windows along optic axis	1800	-
Planar windows along optic axis	700	-
S/N, 1 μsec rise-time[h]	200	100

[a] This is the standard cell shown in Fig. 3.

[b] This cell is similar to that of Fig. 4 in that the sample is contained in a quartz tube with a cylindrical bore. The outside surface is polished flat with a square profile so as to minimize stray light. The values in parentheses refer to a second exchangeable tube with the same outside form but the indicated smaller inner diameter.

[c] The filling volume is about 20% greater.

[d] Normally, condensers in the range 10 to 50 nF and voltages less than 35 KV are used, corresponding to about 30 J. A useful calibrating solution is 10^{-5} M 4-methyl-umbelliferone in 1 M KCl, 0.1 M tris-HCl, pH (20°C) = 7.4 [57]; the fluorescence decreases about 2.4%/°C (λ_{exc} = 365 nm, λ_{em} > 410 nm).

[e] Measured at 10 KHz with 1 M KCl. Thus the resistance of a cell containing a given solution is the cell constant divided by the conductance.

[f] Convection is minimized in these cells because almost the entire contents are heated. Cooling is a complicated physical process, however, and the indicated values are only representative. A linear rate is not implied, and thermal re-equilibration is usually complete within 3 min after a temperature jump.

[g] The ratio is that of the transmitted light intensities of a He-Ne laser beam passed through a cell interposed between two parallel and crossed analyzers, respectively.

[h] These signal-to-noise ratios were achieved with the calibrating solution described in (d).

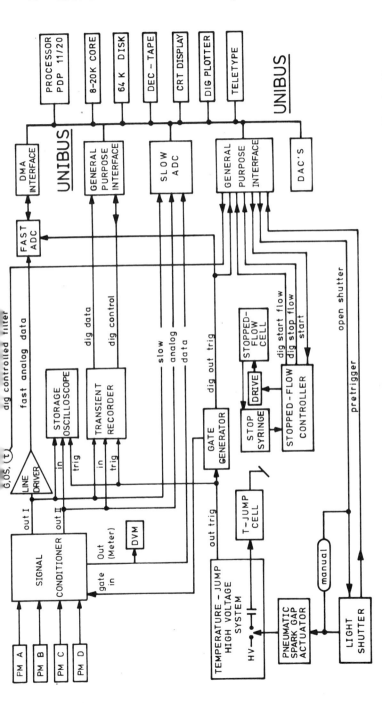

FIG. 5. Electronic components of kinetic system. There are basically two analog paths to the computer, one for fast data over a line driver with adjustable gain (G), offset (OS), and filter rise time (τ) to a 10-bit (± 1 V differential input) 4 MHz ADC (AN-DI 1002 VID, Micro Consultants Ltd, England). The output of the ADC passes into the computer system (Digital Equipment Corp.) through a DMA interface (DR11-B). Slow analog data are collected through a 32-channel multiplexed programable-gain DEC system (AD01-D). The general purpose interfaces (DR11-C) control the flow of digital data and interrupts. The CRT display (Textronix 604 or 613) is operated either in refreshing or storage mode through a DEC subsystem (AA11-D) which also controls other DACs. The digital plotter is a Houston DP-10.

instantaneous signal changes, unwanted faster relaxation processes, or baseline fluctuations. Extensive provisions have been made for computer control of the essential functions of the instrument and automated data acquisition. With the system as shown, analog data can be digitized and stored directly in core with 10-bit resolution and a range of ± 1 V at a present maximum rate of 1.2 MHz (although this rate can be doubled). The system can also operate in the absence of the computer using oscilloscopes or X-Y recorders as the data storage devices.

Examples of kinetic data are given in a later section of the chapter. The sensitivity of the instrument in static measurements was assessed, in one test, by using solutions of L-tryptophan in water. The conditions were: lamp, 200 W Hg-Xe; monochromator, Jarrell-Ash, 280 nm (slit width · dispersion = 7 nm); cutoff filters, Schott WG 305 (3 mm). The photomultiplier response was linear over four decades of concentration, and a solution of 10^{-7} M tryptophan gave a signal about 100% over background luminescence. The absolute sensitivity of the photomultipliers, however, is at least 250 times greater.

C. Design and
Optimization of Experimental Conditions

There are basically two categories of considerations and factors which determine the quality of the data and thus information gained from chemical relaxation experiments. Of primarily quantitative importance are the intensity and stability of the detected light (S/N ratio), the presence or absence of artifacts, the magnitude of the kinetic effects (absolute and relative relaxation amplitudes), and the methods of data acquisition and analysis. Of more qualitative significance is the systematic strategy employed for testing a hypothetical mechanism and for eliminating other reaction schemes. One has recourse to the use of specific conditions under which qualitative distinctions between mechanisms can be made, the selective suppression or enhancement of effects by chemical or spectral manipulations, and the use of multiple parameters and complementary techniques.

The optimization of signal-to-noise ratios has been treated fully elsewhere [39] and is obviously a question intimately related to factors discussed in Table 3 and Fig. 2. Self-absorption in fluorescence, which can pose as an artifact, has been extensively discussed as well [60]. Scattered light can be minimized with horizontally oriented analyzers [61], stray light compensated by suitable offset procedures, and electrode discharges suppressed by careful cell design and sample preparation [39].

In a coupled reaction scheme, the relaxation amplitudes are complicated functions of kinetic, thermodynamic, and spectral constants, as discussed above. It is instructive, however, to consider the example of a simple but important case, the one-step association reaction

$$A + B \underset{k_2}{\overset{k_1}{\rightleftarrows}} C$$

$$nA_o = \bar{c}_A + \bar{c}_C \qquad (65)$$

$$B_o = \bar{c}_B + \bar{c}_C$$

where n is the number of independent and equivalent binding sites per molecule A_o, $\bar{c}_A$ is the concentration of free sites, B is a ligand, A_o and B_o are total initial molecular concentrations, and $K = k_2/k_1 = \bar{c}_A\bar{c}_B/\bar{c}_C$ is the dissociation constant. From Eqs. (50), (49), and (47), expressions for the reciprocal relaxation time, relaxation amplitude, and their product are obtained by inspection:

$$\lambda = rg$$

$$\Delta P = -g^{-1}\Delta\varphi\Delta\ln K \qquad (66)$$

$$\lambda \cdot \Delta P = -r\Delta\varphi\Delta\ln K$$

where

$$r = k_1 \, \bar{c}_A \bar{c}_B = k_2 \bar{c}_C$$

$$g = 1/\bar{c}_A + 1/\bar{c}_B + 1/\bar{c}_C \qquad\qquad (67)$$

$$\Delta\varphi = \varphi_C - \varphi_A - \varphi_B$$

The negative sign in Eq. (66) results from the use of the dissociation constant, as opposed to the convention used in Eq. (2) and in
the derived amplitude expressions. It is seen that the amplitude
depends upon only one concentration-dependent term, g^{-1}, which is
denoted as Γ [Eq. (4)]. A general equation for Γ, normalized to K
and thus applicable for all such systems, is

$$\Gamma/K = \frac{1}{2} \{[1 - 4[\alpha/(1 + \alpha)]^2 \, \beta/(1 + \beta)^2]^{-1/2} - 1\} \qquad (68)$$

where

$$\alpha = (nA_o + B_o)/K \quad \text{and} \quad \beta = A_o/B_o \qquad\qquad (69)$$

This function has the following properties: for any value of α, the
amplitude is maximum for $\beta = 1/n$; for any given B_o, the amplitude is
maximum for $A_o = (B_o + K)/n$; and as $\alpha \to \infty$, Γ/K increases monotonic-
ally and reaches a limiting value

$$\Gamma/K \to 1/(n\beta - 1) \qquad\qquad (70)$$

Thus, for $n\beta - 1$, the amplitude increases without limit, but for all
other β it attains a finite plateau value.

The above considerations apply to the amplitude of a given
single-step process. However, the approach is general and can be
of great value, especially in the initial stages of an investigation.
Relative amplitudes will be further considered below. In general,
of course, the condition for maximal amplitude is only one factor in
the overall experimental strategy. The determination of the rate
constants in the above reaction, for example, in some instances may
be accomplished most easily by measurements at very low values of
β, for which $\lambda \cong k_1 B_o + k_2$. The value of n is not required, but
one is limited to a region of relatively low amplitudes. For the

systematic investigation of the mechanism and the determination of the corresponding parameters ($\Delta\ln K$, ΔH, $\Delta\varphi$, k_1, k_2, K, and n), certain procedures lead to economy of material and easy analysis. Thus, for the above example again, the concentration dependence of the relaxation times and amplitudes are investigated conveniently in experiments involving either titrations (change of β) or dilutions (change in α with invariant β), as further discussed below.

An essential and particularly worthwhile experimental dimension in fluorescence measurements is the array of available chromophores and their potential spectral interrelationships (Table 1). In this connection, fluorescence probes [3] assume special importance.

The selective suppression of relaxation effects can assist in the search for other processes with lower magnitude. This feature is of particular importance in the measurement of relaxation times which are close to each other on the time scale but which may be spectrally differentiable. In fluorescence measurements using polarization optics (Table 3), it is always possible to find instrumental settings under which a given relaxation process can be minimized or eliminated. Similarly, the changes of emission in different parts of the spectrum can be isolated or measured simultaneously by using different filters on photomultipliers A and B (Fig. 2).

Finally, if time resolution is a problem, recourse can be had to smaller condensers and/or to a shunt cell, as already mentioned, in the event the cell conductance cannot be increased by added electrolyte.

D.   Data Acquisition and Analysis

There exists a considerable literature on the use of computers in the acquisition and analysis of kinetic data [63, 24 (Hammes, 1974)]. This discussion will be limited, therefore, to the features of the system developed in our laboratory and named MIDAS (modular

interrogative data acquisition system. Written in assembler lan-
guage and intended for the PDP-11 family of 16-bit computers (DEC),
MIDAS provides flexibility and modularity for a host of fast-to-
medium-speed on-line applications. MIDAS uses the hardware elements
shown in Fig. 5 and consists basically of input-output handling routines, an interrogative monitor named ASK, and a set of conventions
for storing data and calling routines. MIDAS subsystems control
the input of data to the computer, the operation of the kinetic
apparatus and computer peripherals, and the systematic reduction
and evaluation of data.

A large array of methods exists for the analysis of data repre-
sentable by overlapping exponential decays (possibly convoluted
with forcing functions of finite duration). A discussion of this
extensive topic is beyond the scope of this chapter, and reference
is made only to excellent monographs and reviews on data analysis
[64], applied regression analysis [65], digital signal processing
[66], and the methods of (a) moments [67], (b) modulating functions
[68], (c) Laplace transformation [109], and (d) least squares [110].
Several authors have addressed themselves specifically to the analysis of chemical relaxation curves [21,63,69]. The procedure we are
currently following in our laboratory is the systematic evaluation
and comparison of available methods with real and simulated data.
The vast literature on generalized kinetic progress curves, especially with relation to enzymatic studies, will not be treated here.

Assuming that the relaxation spectra have been adequately analyzed, the major problem of evaluation remains, namely, the quantitative interpretation of the data with respect to potential mechanisms and the calculation of the pertinent chemical parameters.
The emphasis usually has been upon the relaxation times [21,23].
Relaxation amplitudes have received more recent attention, particularly in a series of papers by Thusius on the evaluation of data
for a variety of postulated and actual systems with the use of
multiparameter linear or nonlinear regression analysis [26,31,32].

Other publications have dealt with the extraction of thermodynamic constants from amplitude measurements [21-33].

In the following as yet restricted but illustrative and useful form of analysis, stress is put on the combined use of relaxation times and amplitudes. If we consider again the simple association reaction discussed earlier, Eq. (65), a number of simple procedures become evident for the following.

1.   Determination of $\Delta \ln K$

From Eqs. (66) and (67) the product of the reciprocal relaxation time and amplitude can be written as

$$\lambda \cdot \Delta P = -k_2 \overline{c}_C \Delta \varphi \Delta \ln K \tag{71}$$

Equation (71) indicates that $\lambda \cdot \Delta P$ is directly proportional to the concentration of complex C and thus in a sense can be regarded as a kinetic titration parameter. Since Eq. (71) retains its validity even in a multistep mechanism, assuming that the step under consideration equilibrates more rapidly than all others, one has a convenient method for isolating a part of the mechanism which in a static measurement might otherwise be inextractable owing to superposition. If, in addition, complex C is the only species making a spectral contribution (as is often the case for a fluorescence ligand such as anilino-naphthalene sulfonate or ethidium bromide which show large fluorescence quantum yield changes upon binding to macromolecules), then the term $\Delta \varphi \overline{c}_C$ is equal to the function P [Eqs. (53) and (54)]. It thus follows that the relative relaxation amplitude takes the form

$$\Delta P/P = -k_2 \Delta \ln K \cdot \tau \tag{72}$$

where τ is the relaxation time. In fluorescence measurements, $\Delta P/P$ is simply M/S^0, the ratio of the observed (experimental) amplitude to the corrected total signal S^0, Eq. (58). Equation (72) provides a simple linear relationship for obtaining $\Delta \ln K$ (and thus ΔH) by using all data points and a value for k_2 obtained from the analysis in part 2 or a more restricted relaxation time analysis. In a more general version of Eq. (72), provision is made for a finite spectral contribution from the unbound ligand B.

$$\Delta P/(P - \varphi_B B_0) = -k_2 \Delta \ln K \cdot \tau \tag{73}$$

In fluorescence measurements, ΔP and P in Eq. (73) can be replaced directly by M and S^0, respectively, in which case φ_B must be expressed in corresponding units.

2. The Determination of k_1, k_2, and n

An alternative formulation of Eq. (66) is

$$\lambda \cdot M = (\lambda_0 - \lambda) \cdot Z \tag{74}$$

in which

$$\lambda_0 = k_1(nA_0 + B_0) + k_2 \tag{75}$$

and

$$Z = -K/2 \cdot \Delta\varphi\Delta \ln K \tag{76}$$

The quantity λ_0 is the hypothetical reciprocal relaxation time which would be observed for the system in the absence of association, while Z is a constant which contains the thermodynamic and physical parameters (and instrument factors in the case of fluorescence measurements). Three types of experiments are possible: titrations with a fixed B_0 and changing A_0, titrations with a fixed A_0 and a changing B_0, and dilution experiments with fixed ratios of A_0 and B_0. In all cases, Eq. (74) can be applied by rearrangement to a form in which the experimental quantity $\Delta(\lambda \cdot M)/\Delta\gamma$ is related to $\Delta\lambda/\Delta\gamma$, where γ is the appropriate concentration factor relative to a reference condition in each series, and the symbol Δ denotes the algebraic difference in the indicated quantities from those corresponding to the reference state (denoted by starred quantities):

$$B_0^* \text{ fixed, } \gamma = A_0/A_0^*$$

$$\Delta[\lambda \cdot M]/\Delta\gamma = Z[k_1 B_0^* n - \Delta\lambda/\Delta\gamma] \tag{77}$$

A_o^* fixed, $\gamma = B_o/B_o^*$

$$\Delta[\lambda \cdot M]/\Delta\gamma = Z[k_1 A_o^* - \Delta\lambda/\Delta\gamma] \qquad (78)$$

$\gamma = A_o/A_o^* = B_o/B_o^*$

$$\Delta[\lambda \cdot M]\Delta\gamma = Z[k_1(nA_o^* + B_o^*) - \Delta\lambda/\Delta\gamma] \qquad (79)$$

Simple linear plots derived from Eqs. (77)-(79) yield various estimates for k_1, k_2, n, and Z [76]. The latter quantity, when combined
with the results of section 1, gives $\Delta\varphi$.

If one is dealing with a single observed relaxation process in
what is potentially a multistep mechanism, the first essential procedure in data evaluation is the assignment of the relaxation to a
given reaction step or steps. The difficulty in distinguishing between even simple sequential and parallel mechanisms has been
stressed [62]. Even in the case of a two-step sequential reaction
with well-separated relaxation times, differentiating between the
two relaxation processes can be problematical. The combination of
experimental approaches described above, however, can be of great
utility in such cases. Consider the reaction

$$A + B \underset{k_2}{\overset{k_1}{\rightleftarrows}} C \underset{k_4}{\overset{k_3}{\rightleftarrows}} D \qquad (80)$$

where $K_1 = k_2/k_1$, $K_2 = k_3/k_4$, $\varphi_2 = \varphi_D - \varphi_C$, and $\Delta\varphi_1$ is given by
Eq. (67). The formal analyses detailed earlier yield the following
expressions for the two reciprocal relaxation times and amplitudes
under the assumption of large temporal separation:

$$\lambda_I = k_1(\bar{c}_A + \bar{c}_B) + k_2 \qquad (81)$$

$$\lambda_{II} = \frac{k_3(\bar{c}_A + \bar{c}_B)}{K_1 + \bar{c}_A + \bar{c}_B} + k_4 \qquad (82)$$

$$\Delta P_I^o = -\Gamma_1\Delta\varphi_1\Delta\ln K_1 \qquad (83)$$

$$\Delta P_{II}^{O} = \frac{[\Delta\varphi_2 + \Gamma_1\Delta\varphi_1/\bar{c}_C][\Delta\ln K_2 - \Gamma_1\Delta\ln K_1/\bar{c}_C]}{[\Gamma_2^{-1} - \Gamma_1/\bar{c}_C^2]} \tag{84}$$

where

$$\Gamma_1^{-1} = 1/\bar{c}_A + 1/\bar{c}_B + 1/\bar{c}_C \tag{85}$$

$$\Gamma_2^{-1} = 1/\bar{c}_C + 1/\bar{c}_D \tag{86}$$

and the denominator of Eq. (84) by virtue of Eq. (50) is also given by $(\tau_{II}r_2)^{-1} = (\tau_{II}k_3\bar{c}_C)^{-1}$. Related analyses and formulas are found in Refs. [21], [26], and [16].

In a titration experiment at low concentrations relative to K_1, both reciprocal relaxation times increase with concentration, as do the amplitudes which reach a maximum, then decrease. The relaxation parameters for the two processes, however, behave differently at high concentrations.

Titration: A_o fixed; $B_o\uparrow$

$$\lambda_I \to \infty \qquad \lambda_{II} \to k_3 + k_4$$

$$\Gamma_1 \to 0 \qquad \Gamma_2 \to A_o \cdot K_2/(1 + K_2)^2 \tag{87}$$

$$\Delta P_I^O \to 0 \qquad \Delta P_{II}^O \to \Gamma_2\Delta\varphi_2\Delta\ln K_2$$

Dilution: $\beta = A_o/B_o$ fixed; A_o, B_o $\uparrow$

$$\lambda_I \to \infty \qquad\qquad\qquad \lambda_{II} \to k_3 + k_4$$

$$\Gamma_1 \to K_1/[(\beta - 1)(1 + K_2)] \quad \Gamma_2 \to A_o \cdot K_2/(1 + K_2)^2 \tag{88}$$

$$\Delta P_I^O \to -\Gamma_1\Delta\varphi_1\Delta\ln K_1 \qquad \Delta P_{II}^O \to \Gamma_2\Delta\varphi_2\Delta\ln K_2$$

The following is an interpretation of these results. (a) In both instances, the first reciprocal relaxation time increases without limit, but the second achieves a plateau value. (b) In a titration experiment, the first amplitude goes to 0 in the limit, but the second amplitude remains finite. (c) In a dilution experiment, the first amplitude achieves a finite limiting magnitude, but the second increases without limit. It is the latter behavior that can clearly differentiate the two relaxation effects in regions of concentration where the behavior of the relaxation times may be ambiguous. An example is given in a later discussion of actual data.

The above analysis is illustrative of procedures that can be of arbitrary sophistication in the evaluation of experimental data. A general approach is summarized in Fig. 6 in which the interrelationships between the theoretical formulation for a given reaction mechanism and the experimental parameters are shown. For simplicity, dependent reactions are not considered. The overall relaxation time τ^*, introduced by Schwarz [22], is the mean reciprocal relaxation time weighted by the respective amplitudes and is given experimentally by the initial slope of the relaxation curve

$$1/\tau^* = \Sigma\lambda_i \cdot \Delta P_i^O / \Sigma\Delta P_i^O = (\delta P/\delta t)_{t=0}/\Delta P_{tot}^O \tag{89}$$

and in terms of the matrix notation used earlier

$$1/\tau^* = \underline{\Delta\varphi}^T\underline{r}\ \underline{\Delta \ln K}/\Delta P_{tot}^O \tag{90}$$

IV.   APPLICATIONS OF FLUORESCENCE
TEMPERATURE-JUMP AND STOPPED-FLOW TECHNIQUES

This section is concerned with concrete examples of the techniques and instrumentation described above and gives a brief survey of the related literature. Most of the data shown were not acquired with the on-line computer system (Fig. 5) which in more recent work has yielded relaxation curves with much greater precision and accuracy.

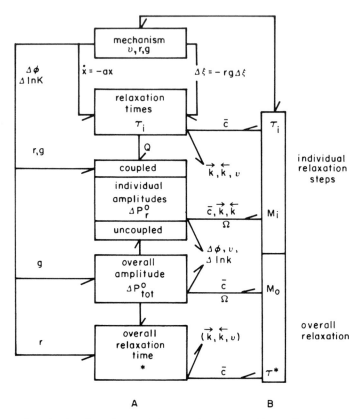

FIG. 6. Analysis of reaction mechanisms with relaxation kinetics.
(A) Theoretical. The matrices placed to the side of the pathway indicate the necessary and sufficient information for the analytical
computation of the desired parameter(s). (B) Experimental data.
See text for details.

A. Binding Equilibrium

The binding of 8-anilino naphthalene-1-sulfonate (ANS) to bovine serum albumin (BSA) is accompanied by characteristic spectral
shifts and an increase in the quantum yield of the dye [70]. There
are five sites which interact, as evidenced by cooperative binding
isotherms [70], depolarization of dye fluorescence due to energy
transfer between molecules liganded to the same protein, quenching
of protein fluorescence with concomitant sensitization of ANS

fluorescence due to heterologous energy transfer [71], and an inter-
dependent orientation of complexed dye molecules [72]. The BSA-ANS
interaction is a good model system owing to its easy availability,
its representative nature in view of the widespread use of ANS as a
fluorescence probe [3], the rich spectral manifestations of the
system, and the suggestion of slow relaxation phenomena in protein
conformation secondary to complex formation as possible elements in
the anomalous binding [70,38].

When examined in the fluorescence temperature-jump apparatus,
the system showed a major relaxation process (Fig. 7) which was concentration dependent and not analyzable in terms of a single relaxation time. The kinetic complexity reinforces the evidence for
multiple reaction steps derived from static measurements and pre-
cludes a simple analysis. The data in Fig. 7 are primarily intended, therefore, as illustrations of the use of many spectroscopic
parameters in the study of such systems. That is, in the same relaxation process occasioned by a $3.6^{\circ}C$ rise in temperature, the
following occurred: The ANS fluorescence decreased (1.7%) as seen
with optical configurations measuring a general signal (Fig. 7A,
Table 3) or more specifically the total emission (Fig. 7B, Table 3);
the polarization of ANS emission was unaffected (Fig. 7C); the absorbance at 365 nm decreased slightly; the ANS fluorescence excited
directly and indirectly at 280 nm decreased (Fig. 7E); and the protein fluorescence increased (Fig. 7F).

Clearly, the ANS-BSA complex is destabilized at higher temperature, and the major effects signaling the change relate to the
local environment of the dye (quantum yield) and to the protein-dye
interaction (energy transfer). The absorbance signal was necessarily
low owing to the inherent limitation in sensitivity under the conditions of Fig. 7 [Eq. (59)] and the use of a wavelength probably
close to an isosbestic point [73]. No polarization signal was observed because the unbound dye is virtually nonfluorescent, and at
a low stoichiometric ratio of ANS and BSA (Fig. 7), the polarization
is a weak function of the extent of binding [71]. The static polar-

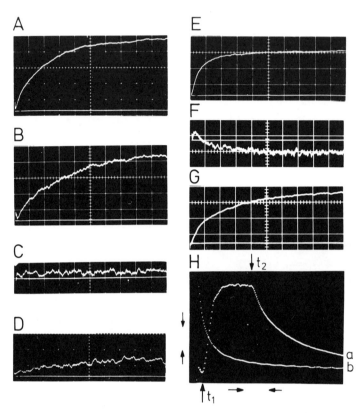

FIG. 7. Kinetic study of the BSA-ANS system. Unless otherwise indicated, the conditions were: buffer (0.1 M KPO$_4$, pH 7), temper- ature (initial: 20°C; ΔT = 3.6°C), concentration ([BSA] = [ANS] = 15 µM), λ_{exc} (365 nm), excitation parameter s (1/2, depolarizer), λ_{em} (>390 nm), time scale (2 msec/division), and the experiment was a temperature-jump using the Cyclops cell (Table 4). Vertical scale is in units/major division and increases in upward direction. (A) Enhanced emission: vertical scale (-0.3%), time scale (5 msec/ division); (B) total emission: excitation (s = 1), analyzers (θ,ψ = 54 deg), vertical scale (-0.4%); (C) polarization: excitation (s = 1), analyzers (θ = 0 deg, ψ = 90 deg), initial polarization (0.26, uncorrected), vertical scale (-0.002%); (D) transmission: vertical scale (0.06%); (E) sensitized fluorescence: λ_{exc} (280 nm), concen- tration (2.4 µM), vertical scale (-1.1%); (F) donor (protein) quench- ing: λ_{exc} (280 nm), λ_{em} (365 nm), concentration (2.4 µM), vertical scale (-0.4%); (G) total emission: microcell (Table 4), concentra- tion (2.4 µM), time scale (5 msec/division), vertical scale (-1.8%); (H) stopped-flow: final concentration (3 µM), vertical scale (10%,

ization value of 0.26 was in good agreement with the data of Weber and Daniel [71]. An experiment using the microcell described in Table 4 is also shown (Fig. 7G); it provides good visual evidence for the spectrum of relaxation times seen with ANS-BSA. A stopped-flow experiment (Fig. 7H) was consistent with the rapid rate of association (8×10^8 M^{-1} sec^{-1}) already reported for this system [74]; the data could be analyzed in terms of a second-order process with $k_{ass} > 10^8$ M^{-1} sec^{-1}. It should be noted that owing to the low concentrations which can be used with the fluorescence temperature-jump method, relaxation times in the range of 1-10 msec are obtained (Fig. 7), even with such a rapidly equilibrating system as this one.

B. Equilibration in the Bound State

As opposed to ANS, there appears to be only one binding site on BSA for 2,6-naphtholsulfonate [9 (1972)]. Upon binding, the emission spectrum of the dye shifts dramatically from a peak at 425 nm, characteristic for the ionized species, to a peak at 360 nm, corresponding to the fully protonated species. No excited-state proton transfer has been observed [9 (1972)].

Preliminary relaxation kinetic data for this system are shown in Fig. 8. One major relaxation process was observed which, in contradistinction to the case with ANS, shows: no concentration dependence of the relaxation time or relative relaxation amplitude over a concentration range of 3 to 22 μM in dye and BSA, an increase in the total fluorescence emission throughout the emission spectrum (Fig. 8A,B) and a large increase in the polarization of

FIG. 7 (Cont'd.). fluorescence). Points t_1 and t_2 represent the start and cessation of flow, respectively, and the time scale for curve a is 5 msec/unit. Curve b represents the data starting from t_1 with a time scale of 40 msec/unit achieved by 8-point linear filtering. In experiments A-G, the instantaneous signal after heating was suppressed electronically, and the data are original oscillograph tracings. H is shown in the form of a computer display.

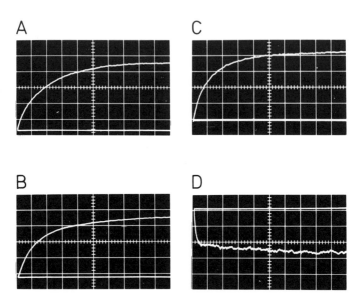

FIG. 8. Relaxation kinetics of BSA and 2,6-naphtholsulfonate.
Original oscillograph tracing of temperature-jump experiments using
the Cyclops cell under the conditions: buffer (0.1 M KPO$_4$, pH 7),
temperature (initial: 20°C, ΔT = 3.6°C); λ$_{exc}$ (313 nm); λ$_{em}$ (> 410
nm with filter KV418 in experiments A and D and > 330 with filter
WG345 in experiments B and C); time scale (50 msec/unit). The conventions for the vertical scale units as in Fig. 7. (A) Total
emission: [BSA] (3 μM), naphtholsulfonate (3.3 μM), vertical scale
(1.8%); (B) total emission: [BSA] (22 μM), vertical scale (2.3%);
(C) polarization: concentrations (as in B), vertical scale (0.03);
(D) polarization: concentrations (as in B), vertical scale (0.016).

emission. The latter effect depended upon the region of the emission
selected for observation. Thus, with a cutoff filter transmitting
light above 340 nm (Fig. 8C), the polarization increased 160% from
a starting value of 0.08, while with a filter transmitting only
above 400 nm (Fig. 8D), with which the recorded emission was 34%
lower, the polarization decreased about 30% from an initial value
of 0.05.

These facts and the absolute value of the relaxation time of
about 50-100 msec (the presence of a small slower component, possibly
due to BSA oligomers, made evaluation difficult) suggest that the

relaxation process is unrelated to the binding step itself but is related more likely to some rearrangement of the complex. Thus, one interpretation of the data is that at the higher temperature, the equilibrium shifts in favor of a species with higher polarization, higher quantum yield, and a relatively blue-shifted emission. Such a trend was seen in measurements of naphtholsulfonate saturated with BSA in the range of 21° to 26°C. Surprisingly high polarization values of up to 0.37 were obtained with narrow band excitation and emission. The low value of the initial polarization seen in the temperature-jump experiments was due to the presence of excess unbound dye and the integrative nature of the detection system. Further work is in progress to elucidate the nature of the relaxation effects and their possible relationship to excited state processes and to protein conformational equilibria, particularly since isomerization reactions in this system have not been reported previously [9 (1972)].

C. Intercalation Reaction

The fluorescence of the dye ethidium bromide (EB) shows a marked enhancement when bound to double-stranded nucleic acid, presumably due to intercalation [75]. While kinetic studies of such interactions have been reported [76,77], they have used exclusively absorbance measurements at relatively high concentrations of dye and nucleic acid. The binding of EB to the synthetic DNA poly (dA-dT) has been studied with the fluorescence temperature-jump method (Figs. 9-12) under low concentration conditions [78], to minimize dye-dye interactions in the free and bound states [75,79].

The data are consistent with a single-step reaction from the unbound to the intercalated species as evidenced by a single relaxation time (Fig. 9), dependencies of the relaxation times and amplitudes on concentration variables which are in accordance with Eqs. (65)-(79) (Figs. 10, 11), the presence of an instantaneous signal

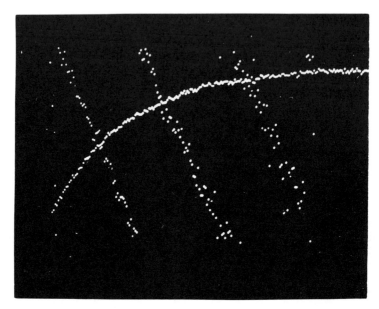

FIG. 9.   Chemical relaxation of poly(dA-dT) and ethidium bromide.
Computer acquired and displayed data of a fluorescence temperature-
jump experiment using the Cyclops cell under the conditions: buffer
(0.1 M KCl, 10 mM KPO$_4$, pH 7.4); ethidium bromide (1 μM); poly(dA-dT)
(2.5 μM nucleotide); temperature (initial: 20°C, ΔT = 4.6°C);
λ_{exc} (303 nm); excitation parameter s (1/2, depolarizer), λ_{em} (> 540
nm). The kinetic curve is in the direction of decreased fluores-
cence. The parallel slanting lines represent the highly magnified
semilog display of the data corresponding to a linear least-squares
analysis yielding a τ of 14.7 msec and an amplitude of -500 mV (about
-10%).

change [Eq. (56)] which is a constant fraction of the total signal
(-1.5%), and a relative amplitude related linearly to the relaxation
time according to Eq. (72) (Fig. 12). It should be added, however,
that the presence in vanishing concentration of a steady-state intermediate between the unbound and intercalated states cannot be excluded.

An analysis of the data yields the following set of parameters
for the system (0.1 M KCl, pH 7, 25°C):

$$k_{on} = 8.5 \pm 0.5 \times 10^6 \ M^{-1} \ sec^{-1}$$

$$k_{off} = 35 \pm 4 \ sec^{-1}$$

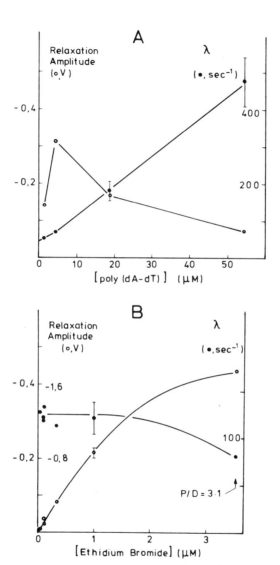

FIG. 10. Relaxation kinetics of poly(dA-dT) and ethidium bromide. Concentration dependence of relaxation amplitudes and reciprocal relaxation times (λ) from fluorescence temperature-jump experiments under conditions similar to those in Fig. 9 except: buffer (0.1 M KCl, 5 mM KPO$_4$, 1 mM 2-mercaptoethanol, pH 7); λ_{em} (>580 nm). (A) Titration of ethidium bromide (0.53 μM) with poly(dA-dT); (B) titration of poly(dA-dT) (11 μM) with ethidium bromide. Each point represents several experiments with the ranges shown with error bars.

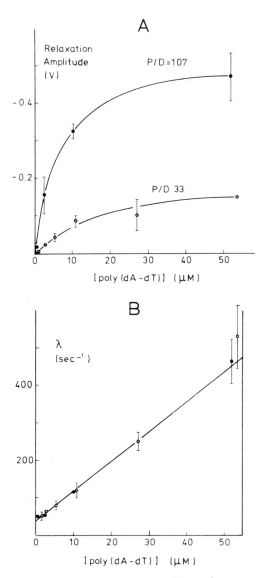

FIG. 11.   Relaxation kinetics of poly(dA-dT) and ethidium bromide
(dilution experiments).   Conditions the same as in Fig. 10.   (A)
Concentration dependence of the relaxation amplitude at two P/D ratios; (B) concentration dependence of the reciprocal relaxation
times corresponding to the experiments in part A.

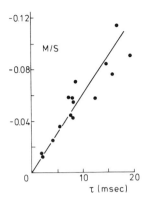

FIG. 12. Determination of ΔH for the reaction of poly(dA-dT) with ethidium bromide. The data from Figs. 10 and 11 are plotted according to Eq. (47), where the ordinate represents the ratio of the experimental relaxation amplitudes to the total initial signal. From the slope one obtains $\Delta \ln K = 0.17$ and $\Delta H = 6.6$ kcal/mole.

$$K_d = 4.1 \pm 0.5 \times 10^{-6} \text{ M}$$

$$\Delta H = 6.6 \pm 1.1 \text{ kcal/mole} \tag{91}$$

The values for k_{on} and K_d are stated with reference to DNA nucleotide content and contain implicitly the value of the stoichiometric coefficient n, the number of bases per binding site [Eq. (65)]. Since virtually all experiments were performed at relatively high P/D ratios, the approaches suggested for the determination of n [Eqs. (77)-(79)] are not valid. However, the plateau values achieved by the relaxation amplitudes in the dilution experiments (Fig. 10) and the single value of the relaxation time at P/D = 3.1 (Fig. 11B) are consistent with a value of n between 0.5 and 1. In view of the degenerate nature of the poly(dA-dT) structure, the expected value for n would be 0.5, that is, one potential binding site per base pair. The kinetic and thermodynamic constants determined from the data are in good agreement with values for DNA obtained by other methods [17,77]. It is interesting that no evidence was found for a second species bound either externally or intercalated, nor was any found for the exchange process reported by Bresloff and Crothers [77]. A comparative fluorescence-absorption study should clarify potential discrepancies. A second concentration-dependent relaxation process was seen with oligonucleotides of (dA-dT), however, and may reflect binding to loops or termini; similar results have been obtained with poly(dG-dC) [80].

It is worthwhile noting that the very different concentration dependencies of the relaxation times and amplitudes in dilution and titration experiments (Figs. 10, 11) not only permit a quantitative evaluation of the system but also provide qualitative support of the proposed mechanism to the exclusion of some other possibilities, e.g., an isomerization reaction [Eqs. (87), (88)].

D. Probe for Membrane Structure

The fluorescent probe ANS has been shown to bind to membrane fragments isolated from the innervated, excitable face of electroplax cells from the electric organ of Electrophorus electricus [81]. The binding of ANS is sensitive to temperature-induced structural rearrangements of the membrane. In the temperature-jump apparatus, the complex of ANS or the related probe TNS (6-p-toluidinonaphthalene-2-sulfonate) with such membranes displayed a relaxation process (Fig. 13), the relaxation time of which was concentration independent (0.50-0.8 msec under standard conditions) but which decreased rapidly with increasing ionic strength. The effect also disappeared reversibly at temperatures greater than 30°C, but was unaffected by neurospecific agents. The relaxation phenomenon has been interpreted in terms of an isomerization reaction between two forms of the bound dye, reflecting features of protein and lipid organization in the membrane which are sensitive to generalized structural transitions [82].

E. Enzymatic Phosphorolysis of RNA

Nucleic acid-protein interactions are particularly amenable to kinetic assay with fluorescence methods because of the availability

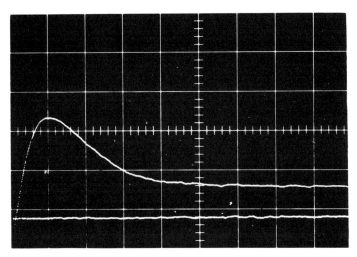

FIG. 13. Relaxation kinetics of the complex of ANS with EME
(excitable membranes from the electric organ of <u>Electrophorus</u> elec-
<u>tricus</u>). Fluorescence temperature-jump experiment with the Cyclops
cell under the conditions: buffer (20% sucrose, 1 mM KPO_4, pH 7);
EME (94 γ/ml protein); ANS (15 μM); temperature (initial: 21°C,
$\Delta T = 4.6°C$, shunt cell for better time resolution); λ_{exc} (365 nm);
excitation parameter s (1); λ_{em} (>410 nm); analyzers and electronics
adjusted for total emission signal; vertical scale (-5%/unit); time
scale (0.2 msec/unit). The initial decrease in fluorescence is due
in part to the instantaneous signal delayed due to the long heating
time and also probably to equilibration of the binding step. The
subsequent relaxation is in the direction of increased fluorescence.

of inherent chromophores as well as base analogs in monomeric and
and polymeric form [4,83] and dyes which bind to the protein or nucleic acid. Since the more specific interactions tend to be very
strong, fluorescence offers the additional advantage of sensitivity
and a multiplicity of parameters usable at low concentration.

Figure 14 shows an example of a stopped-flow experiment in
which the interaction of polynucleotide phosphorylase [84] with the
synthetic, relatively nonfluorescent polymer poly 2-aminopurine [85]
in the presence of inorganic phosphate led to the sequential degradation of the polymer and release of the highly fluorescent mononu-
cleotide. The transient phase is clearly exhibited (Fig. 14b,c) as

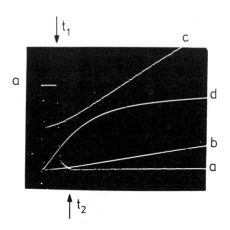

FIG. 14. Degradation of poly-2-aminopurine by polynucleotide
phosphorylase. Computer display of stopped-flow experiment performed under the conditions: buffer (0.1 M tris-HCl, 1 mM $MgCl_2$,
pH 8.3); enzyme (10 nM in one syringe together with 0.5 mg/ml BSA
and 5 mM KPO_4); substrate (0.2 µM in second syringe with buffer);
temperature (20°C); λ_{exc} (303 nm); λ_{em} (channels A and B were
equalized with filter WG 345 in place; filter GG 385 was then in-
stalled on PM A, and A-B recorded after offsetting to 0. This is
thus a method of blank substration). (a) Initial record: t_1 and
t_2 are the initiation and cessation of flow; time scale (170 msec
total); (b) first averaged time line: origin at t_2; time scale
(1.35 sec total); (c) fourfold vertically expanded representation
of (b); (d) second averaged time line: origin at t_2; time scale
(12.8 sec total). Fluorescence increases in the vertical direction.
Data courtesy of H. Lehrach.

is the steady-state of the reaction with the eventual complete hy-
drolysis of the polymer (Fig. 14d). The computer is of invaluable
assistance in the acquisition and analysis of such data.

F. Other Examples of Kinetic Data

A summary of representative studies performed with flow and
relaxation kinetic methods using fluorescence detection is given in
Table 5. A number of additional systems, particularly related to

protein-nucleic acid interactions, are under current investigation in our laboratory.

FUTURE PROSPECTS

The existing kinetic instrumentation is clearly capable of supplying valuable information about a wide range of biological phenomena.   There is room for further development, however, particularly in view of a major criterion for biological studies: economy of material.   Both flow and relaxation methods require highly purified substances, often in amounts much in excess of those required for studies of catalytic function.   Thus it is clear that the emphasis will be on the miniaturization of equipment and a concomitant improvement in time resolution, primarily through the use of lasers as energy [101] and light sources as well as other techniques [102]. The more systematic use of combined flow and relaxation techniques and a multiparameter approach to fluorescence detection will greatly increase the range and versatility of kinetic methods.   Computers already have become indispensable elements in the control of instrumental functions and the acquisition of data.

To utilize fully the potential of computer-oriented systems, more powerful and general theoretical approaches to the analysis of the complex kinetic schemes which realistically describe biological systems are required.

Finally, the various kinetic techniques which are available, both involving fluorescence (Table 2) and other spectroscopic or physical observables, will yield the maximum information if used in a complementary and coordinated fashion.   It can be expected that the range of applications will fully encompass the areas cited in Table 1 and others which may not have been considered.

TABLE 5

Fluorescence Stopped-Flow and Relaxation Kinetic Studies[a]

System	Chromophore	Application[b]	Parameter[c]	Method[d]	Reference
<u>Enzymes</u>					
Dehydrogenases (cofactor, effectors, substrates)					
Cofactor alone	NADH/TPNH	A		TJ	[87]
Glutamate DH	NADH/ANS	B,C,H,I		SF/TJ	[43,54,88]
Alcohol DH	NADH/protein	H,M		SF/TJ	[89]
Homoserine DH I	TPNH/protein	B,H,L		SF/TJ	[90]
Glyceraldehyde phosphate DH	NADH/protein	B,H		SF/TJ	[91]
Lactate DH	NADH/protein	H,L	9	SF/TJ	[92,43]
Malate DH	NADH	H		TJ	[93]
Pyridoxamine pyruvate transaminase (inhibitors, substrates)	Protein	H		TJ	[94]
Lysozyme (saccharides)	Protein	B,H		TJ	[95]
Carboxypeptidase (substrate)	Dansyl	H,F,L	9	SF	[49]
Staphylococcal nuclease (refolding)	Protein	A,B		SF	[96,97]
Carbonic anhydrase (inhibitor)	Dansyl	H		SF	[43]

E. coli RNA, DNA polymerases	Protein/substrates/dyes	B,H-J	4,9	SF/TJ	[35]
Aminoacyl t-RNA synthetases (substrates, cofactors)	Protein/ethidium bromide/Y base/TNS	A-C,H-J,L	4	SF/TJ	[52,55,98]
Polynucleotide phosphorylase (substrate)	2-Aminopurine	H,L		SF	(e)
Flavodoxin apoprotein (flavin)	Flavin	B,H		TJ	[40(3)]
Bacterial luciferase	Light (reaction product)	L		SF	[45]
Myosin ATPase	Protein/formycin	A,B,H,L		SF	[51,104]
Alkaline phosphatase (substrates)	4-Methylumbelliferone	B,H,L		SF	[56,103]
Other Proteins					
E. coli lac repressor (inducer)	Protein	B,C,H		SF	[53]
Antibody (hapten)	Fluorescein/protein	B,H	4	SF/TJ	[48,50,99,113]
Bovine serum albumin (bilirubin, dyes)	Bilirubin/ANS/naphtholsulfonate	B,I	4,9	SF/TJ	[43,38,74](e)
Concanavalin A (saccharides)	4-Methylumbelliferyl-mannoside	B,C,I		SF/TJ	[106]
Globin	Porphyrin	H		SF	[74]
Nucleic Acids					
Nucleic acids (dyes)	Ethidium bromide/proflavin	I	4	SF/TJ	[16,26(4)39, 80](f)

TABLE 5 (continued)

System	Chromophore	Application[b]	Parameter[c]	Method[d]	Reference
Membranes					
Mitochondrial membranes (energy coupling)	ANS/ethidium brcmide	I		SF	[100(1)]
Excitable membranes (probes)	ANS/TNS/ethidium brcmide	I		TJ	[82][e]
Lipid model membranes (phase transitions)	ANS	I		TJ	[100(2,3)]
Phospholipid membranes (ions, carriers, probes)	ANS	I		TJ	[100(4)]
Chromaffin granules	Diphenyl-hexatriene	I		TJ	[114]

[a]General reviews on the biological applications of rapid kinetics are given in Refs. 86, 33 (first entry), and 107.

[b]The classification is that of Table 1. The assignments are necessarily somewhat arbitrary. For example, classes D-G are to be understood for practically every system cited, as is also M, of course.

[c]The parameters are given in Table 1. As in (b), the choice is somewhat arbitrary. Thus the quantum yield (1) and emission spectra (3) are probably affected in most cases and are therefore omitted for clarity.

[d]SF = stopped-flow; TJ = temperature jump.

[e]The numbers in parentheses are the respective entry in the given reference.

[e]This chapter.

ACKNOWLEDGMENTS

I thank my wife, Dr. Donna Jovin, for enormous assistance in the preparation of this manuscript; Dr. Darwin Thusius for innumerable stimulating discussions and his generous supply of unpublished manuscripts; Dr. Robert Clegg for critically reviewing the theoretical discussions; Dr. Hans Lehrach for unpublished results; the many people who contributed to the design and construction of the instrumentation: Dr. Rudolf Rigler, Dr. Roland Rabl, Dr. Hans Lehrach, Mr. Wolfgang Simm, and Mr. Hermann Bründl; and Mr. George Striker for the implementation of the computer system.

NOTES

1. D. M. Hercules (ed.), Fluorescence and Phosphorescence Analysis, Wiley-Interscience, New York, 1966; R. S. Becker, Theory and Interpretation of Fluorescence and Phosphorescence, Wiley-Interscience, New York, 1969; A. J. Pesce, C.-G. Rosén, and T. L. Pasby (eds.), Fluorescence Spectroscopy, Dekker, New York, 1971; C. A. Parker, Photoluminescence of Solutions, Elsevier, Amsterdam, 1968; T. Förster, Fluoreszenz organischer Verbindungen, Vandenhoeck & Ruprecht, Göttingen, 1951 (in German); L. Brand and B. Witholt, in Methods in Enzymology (C. H. W. Hirs, ed.), Vol. 11, Academic, New York, 1967, pp. 776-856; A. Weller, in Progress in Reaction Kinetics (G. Porter, ed.), Vol. 1, Pergamon, Oxford, 1961, pp. 187-214.

2. R. F. Steiner and I. Weinryb (eds.), Excited States of Proteins and Nucleic Acids, Macmillan, London, 1971; L. Stryer, Science, 162, 526 (1968); S. V. Konev, Fluorescence and Phosphorescence of Proteins and Nucleic Acids (S. Udenfriend, transl.), Plenum, New York, 1967; G. M. Barenboim, A. N. Domanskii, and K. K. Turoverov, Luminescence of Biopolymers and Cells (R. F. Chen, transl.), Plenum, New York, 1969; C. R. Cantor and T. Tao, in Procedures in Nucleic Acid Research (G. L. Cantoni and D. R. Davies, eds.), Vol. 2, Harper and Row, New York, 1971, pp. 31-93.

3. B. Chance, C.-P. Lee, and J. K. Blasie (eds.), Probes of Structure and Function of Macromolecules and Membranes, Vol. 1, Academic, New York, 1971; B. Chance, T. Yonetani, and A. S.

Mildvan (eds.), Probes of Structure and Function of Macromolecules and Membranes, Vol. 2, Academic, New York, 1971; L. Brand
and J. R. Gohlke, Ann. Rev. Biochem., 41, 843 (1972); G. M.
Edelman and W. O. McClure, Accounts Chem. Res., 1, 65 (1968);
G. K. Radda, Curr. Top. Bioenerg., 4, 81 (1971); L. Stryer, in
Molecular Properties of Drug Receptors (R. Porter and M. O'Con-
nor, eds.), Churchill, London, 1970, pp. 133-150; M. E. Kirtley
and D. E. Koshland, Jr., in Methods in Enzymology (C. H. W. Hirs
and S. N. Timascheff, eds.), Part C, Vol. 26, Academic, New York,
1972, pp. 578-601; G. K. Radda and J. Vanderkooi, Biochim. Biophys. Acta, 265, 509 (1972).

4. G. Weber, Ann. Rev. Biophys. Bioenerg., 1, 553 (1972).

5. G. Weber, Advan. Protein Chem., 8, 415 (1953); G. Weber, in
    Fluorescence and Phosphorescence Analysis (D. M. Hercules, ed.),
    Interscience, New York, 1966, pp. 217-239; I. Z. Steinberg, in
    Ref. 112, Vol. 1, Chap. 3; S. A. Levison, in Ref. 112, Vol. 1,
    Chap. 7; G. Weber, in Ref. 111, pp. 5-13.

6. H. Schlessinger and I. Z. Steinberg, Proc. Natl. Acad. Sci.,
    69, 769 (1972); I. Z. Steinberg, in Ref. 112, Vol. 1, Chap. 3.

7. W. R. Ware, in Creation and Detection of the Excited State
    (A. A. Lamola, ed.), Part A, Vol. 1, Marcel Dekker, Inc., New
    York, 1971, pp. 213-302; C. Lewis, W. R. Ware, L. J. Doemeny,
    and T. L. Nemzek, Rev. Sci. Instr., 44, 107 (1973); I. Isenberg,
    in Ref. 112, Vol. 1, Chap. 2. See also Refs. 67, 68, 109, 110,
    111.

8. J. Yguerabide, in Methods in Enzymology (C. H. W. Hirs and S. N.
    Timascheff, eds.), Part C, Vol. 26, Academic, New York, 1972,
    pp. 498-578; G. Weber, J. Chem. Phys., 55, 2399 (1971); G. G.
    Belford, R. L. Belford, and G. Weber, Proc. Natl. Acad. Sci., 69,
    1392 (1972); G. Weber, in Methods in Enzymology (K. Kustin, ed.),
    Vol. 16, Academic, New York, 1969, pp. 380-394; M. Ehrenberg and
    R. Rigler, Chem. Phys. Letters, 14, 539 (1972); Ph. Wahl, in Ref.
    112, Vol. 1, Chap. 1.

9. M. R. Loken, J. W. Hayes, J. R. Gohlke, and L. Brand, Biochem-
    istry, 11, 4779 (1972); L. Brand and J. R. Gohlke, J. Biol.
    Chem., 246, 2317 (1971); A. Grinvald and I. Z. Steinberg, Bio-
    chem., 13, 5170 (1974).

10. R. E. Dale and J. Eisinger, in Ref. 112, Vol. 1, Chap. 4; I. Z.
    Steinberg, Ann. Rev. Biochem., 40, 83 (1971); A. Grinvald, E.
    Hass, and I. Z. Steinberg, Proc. Natl. Acad. Sci., 69, 2273
    (1972); P. J. Wagner, in Creation and Detection of the Excited
    State (A. A. Lamola, ed.), Part A, Vol. 1, Marcel Dekker, Inc.,
    New York, 1971, pp. 173-212; L. Stryer, Radiation Res., supp.
    2, 432 (1960); R. B. Gennis and C. R. Cantor, Biochemistry, 11,

2509 (1972); C. R. Cantor and P. Pechukas, Proc. Natl. Acad. Sci., 68, 2099 (1971); P. Schiller, in Ref. 112, Vol. 1, Chap. 5.

11. W. C. Galley and L. Stryer, Biochemistry, 8, 1831 (1969).

12. S. Claesson (ed.), Fast Reactions and Primary Processes in Chemical Kinetics, Almqvist and Wiksell, Stockholm, 1967.

13. C. A. Parker, in Ref. 12, pp. 325-331; W. C. Galley and L. Stryer, Proc. Natl. Acad. Sci., 60, 108 (1968).

14. B. Chance, R. H. Eisenhardt, Q. H. Gibson, and K. K. Lonberg-Holm (eds.), Rapid Mixing and Sampling Techniques in Biochemistry, Academic, New York, 1964; F. J. W. Roughton and B. Chance, in Technique of Organic Chemistry (S. L. Friess, E. S. Lewis, and A. Weissberger, eds.), Part II, Vol. 8, 2nd ed., Wiley-Interscience, New York, 1963, pp. 703-792; B. Chance, in Ref. 34, pp. 5-62.

15. K. Kustin (ed.), Methods in Enzymology, Vol. 16, Academic, New York, 1969.

16. R. Rigler and M. Ehrenberg, Quart. Rev. Biophys., 6, 139 (1973).

17. E. L. Elson and D. Magde, Biopolymers, 13, 1 (1974); D. Magde, E. L. Elson, and W. W. Webb, Biopolymers, 13, 29 (1974).

18. F. W. Willets, in Progress in Reaction Kinetics (K. R. Jennings and R. B. Cundall, eds.), Part I, Vol. 6, Pergamon, Oxford, 1971, pp. 51-74.

19. P. M. Rentzepis and M. R. Topp, Trans. N. Y. Acad. Sci., Ser. II, 33, 289 (1971); G. E. Busch, R. P. Jones, and P. M. Rentzepis, Chem. Phys. Letters, 18, 178 (1973).

20. G. Porter, in Techniques of Organic Chemistry (S. L. Friess, E. S. Lewis, and A. Weissberger, eds.), Part II, Vol. 8, 2nd ed., Wiley-Interscience, New York, 1963, pp. 1055-1106; G. Porter, in Ref. 12, pp. 141-164; G. Porter and M. A. West, in Ref. 34, pp. 367-462; R. G. W. Norrish, in Ref. 12, pp. 33-70.

21. M. Eigen and L. De Maeyer, in Techniques of Organic Chemistry (S. L. Friess, E. S. Lewis, and A. Weissberger, eds.), Part II, Vol. 8, 2nd ed., Wiley-Interscience, New York, 1963, pp. 895-1054; M. Eigen and L. De Maeyer, in Ref. 34, pp. 63-146.

22. G. Schwarz, Rev. of Modern Phys., 40, 206 (1968); J. Meixner, Kolloid-Z., 134, 3 (1953); R. Alberty, G. Yagil, W. F. Diven, and M. Takahashi, Acta Chem. Scand., 17, 34 (1963); H. J. G. Hayman, Trans. Faraday Soc., 65, 2918 (1969); H. J. G. Hayman,

Trans. Faraday Soc., 66, 1402 (1970); H. J. G. Hayman, Trans.
Faraday Soc., 67, 3240 (1971); H. J. G. Hayman, Israel J. Chem.,
9, 419 (1971); H. J. G. Hayman, Israel J. Chem., 11, 489 (1972);
G. M. Loudon and D. E. Koshland, Jr., Biochemistry, 11, 229
(1972); K. Kustin, D. Shear, and D. Kleitman, J. Theoret. Biol.,
9, 186 (1965).

23. G. W. Castellan, Ber. Bunsenges. Phys. Chem., 67, 898 (1963).

24. G. G. Hammes and P. R. Schimmel, in The Enzymes (P. D. Boyer,
     ed.), Vol. 2, 3rd ed., Academic, New York, 1970, pp. 67-114;
     G. G. Hammes and P. R. Schimmel, J. Phys. Chem., 70, 2319
     (1966); G. G. Hammes and P. R. Schimmel, J. Phys. Chem., 71,
     917 (1967); G. G. Hammes, in Ref. 34, pp. 147-185.

25. G. H. Czerlinski, Chemical Relaxation, Arnold, London, 1966;
     G. H. Czerlinski, in Theoretical and Experimental Biophysics
     (A. Cole, ed.), Vol. 2, Marcel Dekker, Inc., New York, 1969,
     pp. 69-157.

26. F. Guillain and D. Thusius, J. Amer. Chem. Soc., 92, 5534
     (1970); D. Thusius, J. Amer. Chem. Soc., 94, 356 (1972); D.
     Thusius, Biochimie, 55, 277 (1973); D. Thusius, G. Foucault,
     and F. Guillain, in Dynamic Aspects of Conformation Changes in
     Biological Macromolecules (C. Sadron, ed.), Reidel, Dordrecht,
     Holland, 1973, pp. 271-284.

27. K. Kustin (ed.), Methods in Enzymology, Vol. 16, Academic, New
     York, 1969; E. F. Caldin, Fast Reactions in Solution, Black-
     well, Oxford, 1964; H. Gutfreund, Enzymes: Physical Principles,
     Wiley-Interscience, London, 1972; A. F. Yapel, Jr. and R. Lumry,
     in Methods in Biochemical Analysis (D. Glick, ed.), Vol. 20,
     Wiley, New York, 1971, pp. 169-350; D. N. Hague, Fast Reactions,
     Wiley-Interscience, New York, 1971; B. H. Havsteen, in Physical
     Principles and Techniques of Protein Chemistry (S. J. Leach,
     ed.), Part A, Academic, New York, 1969, pp. 245-289.

28. R. Winkler, Ph.D. Thesis, Göttingen-Wien, 1969; M. Eigen, in
     Probes of Structure and Function of Macromolecules and Mem-
     branes (B. Chance, C.-P. Lee, and J. K. Blasie, eds.), Vol. 1,
     Academic, New York, 1971, pp. 535-538.

29. G. Czerlinski, J. Theoret. Biol., 17, 343 (1967).

30. P. R. Schimmel, J. Chem. Phys., 54, 4136 (1971).

31. D. Thusius, in Chemical and Biological Applications of Relaxa-
     tion Spectrometry (E. Wyn-Jones, ed.), Reidel, Dordrecht, Hol-
     land, in press.

32. D. Thusius, P. Dessen, and J. M. Jallon, J. Mol. Biol., 92, 413
     (1975); D. Thusius, J. Mol. Biol., 94, 367 (1975).

33. P. B. Chock, Biochimie, 53, 161 (1971); P. B. Chock, Proc. Natl. Acad. Sci., 69, 1939 (1972).

34. A. Weissberger (ed.), Techniques of Chemistry, Part II, Vol. 6, 3rd ed. (G. G. Hammes, ed.), Wiley-Interscience, New York, 1974.

35. T. M. Jovin and K. H. Scheit, Proc. 9th Meeting Fed. Europ. Biochem. Soc. (in press); T. M. Jovin, in Ref. 107.

36. M. L. Applebury, D. M. Zuckerman, A. A. Lamola, and T. M. Jovin, Biochem., 13, 3448 (1974).

37. H. Strehlow, personal communication.

38. H. Nakatani, M. Haga, and K. Hiromi, FEBS Letters, 43, 293 (1974).

39. R. Rigler, C.-R. Rabl, and T. M. Jovin, Rev. Sci. Instr., 45, 580 (1974).

40. G. Czerlinski, Rev. Sci. Instr., 33, 1184 (1962); B. Chance, B. Schoener, and D. DeVault, Science, 144, 561 (1964); B. G. Barman and G. Tollin, Biochemistry, 11, 4746 (1972).

41. R. L. Berger, B. Balko, W. Borcherdt, and W. Friauf, Rev. Sci. Instr., 39, 486 (1968).

42. K. Hiromi, S. Ono, S. Itoh, and T. Nagamura, J. Biochem. (Tokyo), 64, 897 (1968).

43. R. F. Chen, A. N. Schechter, and R. L. Berger, Anal. Biochem., 29, 68 (1969).

44. B. Hess, H. Kleinhans, and H. Schlüter, Z. Physiol. Chem., 351, 515 (1970).

45. J. W. Hastings and Q. H. Gibson, J. Biol. Chem., 238, 2537 (1963).

46. Q. H. Gibson, J. W. Hastings, G. Weber, W. Duane, and J. Massa, in Flavins and Flavoproteins (E. C. Slater, ed.), American Elsevier, New York, 1966, p. 341.

47. B. Chance, D. DeVault, V. Legallais, L. Mela, and T. Yonetani, in Ref. 12, pp. 437-468.

48. L. A. Day, J. M. Sturtevant, and S. J. Singer, Ann. N. Y. Acad. Sci., 103, 611 (1963).

49. D. S. Auld, S. A. Latt, and B. L. Vallee, Biochemistry, 11, 4994 (1972).

50. S. A. Levison, A. J. Portmann, F. Kierszenbaum, and W. B. Dandliker, Biochem. Biophys. Res. Commun., 43, 258 (1971).

51. C. R. Bagshaw, J. F. Eccleston, D. R. Trentham, D. W. Yates, and R. S. Goody, Cold Spring Harbor Sym. Quant. Biol., 37, 127 (1972).

52. E. Holler and M. Calvin, Biochemistry, 11, 3741 (1972).

53. S. L. Laiken, C. A. Gross, and P. H. v. Hippel, J. Mol. Biol., 66, 143 (1972).

54. J. M. Jallon, A. di Franco, and M. Iwatsubo, Eur. J. Biochem., 13, 428 (1970).

55. A. Pingoud, D. Riesner, D. Boehme, and G. Maass, FEBS Letters, 30, 1 (1973).

56. H. N. Fernley and S. Bisaz, Biochem. J., 107, 279 (1968).

57. R. F. Chen, Anal. Letters, 1, 423 (1968).

58. M. Eigen, in Ref. 12, pp. 333-369.

59. Q. H. Gibson and L. Milnes, Biochem. J., 91, 161 (1964).

60. M. Ehrenberg, E. Cronvall, and R. Rigler, FEBS Letters, 18, 199 (1971); H. Braunsberg, and S. B. Osborn, Anal. Chim. Acta, 6, 84 (1952); D. M. Hercules, Anal. Chem., 38 (12), 29A (1966).

61. R. F. Chen, Anal. Biochem., 14, 497 (1966).

62. R. O. Viale, J. Theor. Biol., 31, 501 (1971).

63. D. Garfinkel, L. Garfinkel, M. Pring, S. B. Green, and B. Chance, Ann. Rev. Biochem., 39, 473 (1970); B. G. Willis, J. A. Bittikofer, H. L. Pardue, and D. W. Margerum, Anal. Chem., 42, 1340 (1970); R. J. DeSa and Q. H. Gibson, Comput. Biomed. Res., 2, 494 (1969); M. L. Johnson and T. M. Schuster, Biophys. Chem., 2, 32 (1974); M. Krizan and H. Strehlow, Chem. Instrum., 5, 99 (1974); Y. Bard and L. Lapidus, Catalysis Rev., 2, 67 (1968); T. R. Crossley and M. A. Slifkin, in Progress in Reaction Kinetics (G. Porter, ed.), Vol. 5, Pergamon, Oxford, 1970, pp. 409-435; H.-J. Wieker, K.-J. Johannes, and B. Hess, FEBS Letters, 8, 178 (1970).

64. M. E. Magar, Data Analysis in Biochemistry and Biophysics, Academic, New York, 1972.

65. N. R. Draper and H. Smith, Applied Regression Analysis, Wiley, New York, 1966.

66. B. Gold and C. M. Rader, Digital Processing of Signals, McGraw-Hill, New York, 1969; A. D. Whalen, Detection of Signals in Noise, Academic, New York, 1971.

67. I. Isenberg, R. D. Dyson, and R. Hanson, Biophys. J., 13, 1090 (1973); I. Isenberg, Biochemical Fluorescence: Concepts (R. F. Chen and H. Edelhoch, eds.), Vol. 1, Chap. 2, Marcel Dekker, Inc., New York, 1975.

68. B. Valeur and J. Moirez, J. Chim. Phys., 70, 500 (1973); G. Striker, to be published.

69. H. Strehlow and J. Jen, Chem. Instrumentation, 3, 47 (1971); F. Eggers, Nature Phys. Sci., 229, 89 (1971); G. W. Hoffman, Ph.D. Thesis, Technical University, Braunschweig, 1972; J. Victor, D. Haselkorn, and I. Pecht, Computers and Biomed. Res., 6, 121 (1973).

70. E. Daniel and G. Weber, Biochemistry, 5, 1893 (1966).

71. G. Weber and E. Daniel, Biochemistry, 5, 1900 (1966).

72. B. Witholt and L. Brand, in Ref. 111, pp. 283-290.

73. L. Stryer, J. Mol. Biol., 13, 482 (1965).

74. Q. H. Gibson and E. Antonini, in Hemes and Hemoproteins (B. Chance, R. W. Estabrook, and T. Yonetani, eds.), Academic, New York, 1966, pp. 67-78.

75. J.-B. LePecq and C. Paoletti, J. Mol. Biol., 27, 87 (1967); C.-C. Tsai, S. C. Jain, and H. M. Sobell, Proc. Natl. Acad. Sci., 72, 628 (1975); D. Lang, Phil. Trans. Roy. Soc. London, B261, 151 (1971); J.-B. LePecq, in Methods of Biochemical Analysis (D. Glick, ed.), Vol. 20, Wiley, New York, 1971, pp. 41-86; Ph. Wahl, J. Paoletti, and J.-B. LePecq, Proc. Natl. Acad. Sci., 65, 417 (1970).

76. R. Bittman, J. Mol. Biol., 46, 251 (1969); T. R. Tritton and S. C. Mohr, Biochem. Biophys. Res. Commun., 45, 1240 (1971); T. R. Tritton and S. C. Mohr, Biochemistry, 12, 905 (1973).

77. L. Bresloff and D. M. Crothers, J. Mol. Biol., in press.

78. R. Mikulak and T. Jovin, unpublished results.

79. D. M. Crothers, Biopolymers, 6, 575 (1968).

80. F. M. Pohl, T. M. Jovin, W. Baehr, and J. J. Holbrook, Proc. Natl. Acad. Sci., 69, 3805 (1972).

81. M. Kasai, J.-P. Changeux, and L. Monnerie, Biochem. Biophys. Res. Commun., 36, 420 (1969).

82. T. Jovin, L. Langlotz, and J.-P. Changeux, in preparation.

83. D. Ward, E. Reich, and L. Stryer, J. Biol. Chem., 244, 1228 (1969).

84. T. Godefroy-Colburn and M. Grunberg-Manago, in The Enzymes (P. D. Boyer, ed.), Vol. 7, 3rd ed., Academic, New York, 1972, pp. 533-574.

85. A. Wacker, E. Lodemann, K. Gauri, and P. Chandra, J. Mol. Biol., 18, 382 (1966).

86. M. Eigen and J. S. Johnson, Ann. Rev. Phys. Chem., 11, 307 (1960); M. Eigen and G. G. Hammes, Adv. Enz., 25, 1 (1963); M. Eigen and K. Kustin, ICSU Review, 5, 97 (1963); G. G. Hammes, Accounts Chem. Res., 1, 321 (1968); W. W. Cleland, Ann. Rev. Biochem., 36, 77 (1967); G. G. Hammes, Advan. Prot. Chem., 23, 1 (1968); M. Eigen, Quart. Rev. Biophys., 1, 3 (1968); Q. H. Gibson, in Methods in Enzymology (K. Kustin, ed.), Vol. 16, Academic, New York, pp. 187-228; G. G. Hammes and C.-W. Wu, Science, 172, 1205 (1971); A. N. Schechter, Science, 170, 273 (1970); H. Gutfreund, Ann. Rev. Biochem., 40, 315 (1971); J. F. Kirsch, Ann. Rev. Biochem., 42, 205 (1973).

87. G. Czerlinski and F. Hommes, Biochim. Biophys. Acta, 79, 46 (1964).

88. A. D. B. Malcolm, Eur. J. Biochem., 27, 453 (1972).

89. G. H. Czerlinski, Biochim. Biophys. Acta, 64, 199 (1963); H. Theorell, A. Ehrenberg, and C. de Zalenski, Biochem. Biophys. Res. Commun., 27, 309 (1967); G. Geraci and Q. H. Gibson, J. Biol. Chem., 242, 4275 (1967).

90. J. Janin and M. Iwatsubo, Eur. J. Biochem., 11, 530 (1969); H. J. Bright, in Probes of Structure and Function of Macromolecules (B. Chance, T. Yonetani, and A. S. Mildvan, eds.), Vol. 2, Academic, New York, 1971, pp. 123-127.

91. G. von Ellenrieder, K. Kirschner, and I. Schuster, Eur. J. Biochem., 26, 220 (1972).

92. G. H. Czerlinski and G. Schreck, J. Biol. Chem., 239, 913 (1964); H. d'A. Heck, J. Biol. Chem., 244, 4375 (1969); J. J. Holbrook and H. Gutfreund, FEBS Letters, 31, 157 (1973).

93. G. H. Czerlinski and G. Schreck, Biochemistry, 3, 89 (1964); E. J. del Rosario and G. G. Hammes, Biochemistry, 10, 716 (1971).

94. J. F. Kirsch, H. Winkler, and K. Sundquist, 9th International Congress of Biochemistry, Stockholm, July, 1973.

95. I. Pecht, V. I. Teichberg, and N. Sharon, FEBS Letters, 10, 241 (1970).

96. H. F. Epstein, A. N. Schechter, R. F. Chen, and C. B. Anfinsen, J. Mol. Biol., 60, 499 (1971).

97. R. F. Chen, Abstr. IUPAB Sym., "Relaxation Methods in Molecular Biology," Copenhagen, 1972.

98. G. Krauss, R. Römer, D. Riesner, and G. Maass, FEBS Letters, 30, 6 (1973); R. Rigler, E. Cronvall, R. Hirsch, U. Pachmann, and H. G. Zachau, FEBS Letters, 11, 320 (1970).

99. I. Pecht, D. Givol, and M. Sela, J. Mol. Biol., 68, 241 (1972).

100. B. Chance, Proc. Natl. Acad. Sci., 67, 560 (1971); U. Fischer, Diplomarbeit, Technical University of Munich, Munich, W. Germany, 1973 (in German); H. Träuble, Naturwissens., 58, 277 (1971); D. H. Haynes, 8th Tutzing - Symposium of DECHEMA, "Technische Biochemie," March, 1972, Verlag Chemie, Weinheim, W. Germany, 1973, pp. 119-133.

101. R. Rigler, A. Jost, and L. de Maeyer, Exptl. Cell. Res., 62, 197 (1970); D. H. Turner, G. W. Flynn, N. Sutin, and J. V. Beitz, J. Amer. Chem. Soc., 94, 1554 (1972).

102. G. W. Hoffman, Rev. Sci. Instr., 42, 1643 (1971).

103. H. N. Fernley and P. G. Walker, Biochem. J., 111, 187 (1969).

104. C. R. Bagshaw and D. R. Trentham, Biochem. J., 133, 323 (1973).

105. R. M. Noyes, in Ref. 34, pp. 343-365.

106. R. M. Clegg, F. G. Loontiens, and T. M. Jovin, to be published.

107. R. Rigler and I. Pecht (eds.), Chemical Relaxation in Molecular Biology, Springer-Verlag (in press).

108. C.-W. Wu and F. Y.-H. Wu, Biochemistry, 13, 2573 (1974).

109. A. Gafni, R. L. Modlin, and L. Brand, Biophys. J., 15, 263 (1975).

110. A. Grinvald and I. Z. Steinberg, Anal. Biochem., 59, 583 (1974).

111. A. A. Thaer and M. Sernetz (eds.), Fluorescence Techniques in Cell Biology, Springer, Berlin, 1973.

112. R. F. Chen and H. Edelhoch (eds.), Biochemical Fluorescence: Concepts, Vols. 1 and 2, Marcel Dekker, Inc., New York, 1975.

113. I. Pecht, D. Haselkorn, and S. Friedman, FEBS Letters, 24, 331 (1972); D. Haselkorn, S. Friedman, D. Givol, and I. Pecht, Biochemistry, 13, 2210 (1974).

114. K. Rosenheck, P. Lindner, and I. Pecht, J. Memb. Biol., 20, 1 (1975).

Chapter 7

FLUORESCENCE POLARIZATION KINETIC
STUDIES OF MACROMOLECULAR REACTIONS

Stuart A. Levison[†]
Department of Biochemistry
Scripps Clinic and Research Foundation
La Jolla, California

I. INTRODUCTION

Fluorescence polarization and intensity measurements provide a
powerful means by which macromolecular association reactions can be
studied. During the past few years, these fluorometric techniques
have been applied by this laboratory to rate studies involving antigen-antibody [1 to 4], hapten-antihapten [5,6], and enzyme-substrate
systems [7 to 9]. This article describes the kinetic approaches, as

[†]The author is an Awardee of the National Institute of Health
Special Fellowship (1F03-AM-42-568-01).

as well as some of the more significant results, obtained in these
investigations.

Fluorescence polarization in equilibrium studies was first
utilized by Laurence [10] in dye-protein binding and by Steiner [11]
in protein-protein and protein-nucleic acid associations. Further
important contributions in both the theoretical and experimental
uses of fluorescence polarization and intensity techniques were made
by Dandliker and co-workers in studies concerning both the equilibria
[5,5a,12 to 16] and kinetics [1 to 4,6,16a] of antigen-antibody reactions. A simple and yet general approach has emerged [17], based
on utilizing the inherent sensitivity of fluorescence measurements
in monitoring the extent of reaction as a fluorescent reactant, $\mathfrak{F}$,
combines with its macromolecular partner, $\mathfrak{R}$:

$$\mathfrak{F} + \mathfrak{R} \rightleftarrows \mathfrak{F} - \mathfrak{R} \tag{1}$$

The investigator can choose to follow changes in the fluorescence
polarization and/or the fluorescence intensity. If the reactants
do not have natural fluorescence, as in the case of many antigen-
antibody systems, one of the reactants (such as the antigen, ovalbumin) can be covalently labeled with a small fluorescent tag, e.g.,
fluorescein [18]. An increase in the fluorescence polarization of
$\mathfrak{F}$ usually occurs during combination with $\mathfrak{R}$, even if there are no
concomitant changes in the fluorescence intensity. This is because
the polarization increase reflects a slowing down of the rotary motion of the smaller ligand, $\mathfrak{F}$, when it becomes attached to the
larger species, $\mathfrak{R}$. These methods of following macromolecular reactions can be utilized in the static mode to study slow kinetic processes (10 sec or greater) as well as equilibrium processes, or they
can be combined with stopped-flow and temperature-jump methods to
study the more rapid rate processes in the milli- and microsecond
range. Concentrations which can be readily detected with the instrumentation available in this laboratory range from 10^{-6} to 10^{-12}
M in the static mode (time constant, 5 sec) and 10^{-9} M in the stopped-
flow mode (time constant, 1 msec) for a hapten or tag such as fluo-

rescein [5,6]. For fluorescent substrates such as NADPH, 10^{-7} to 10^{-8} M can be readily measured [9].

II. INSTRUMENTATION

Equilibrium fluorescence polarization and intensity measurements as well as "slow" kinetic measurements were determined in a
direct readout "polarometer" [5,19] capable of measuring both the
degree of fluorescence polarization and the fluorescence intensity
of a solution. Polarometer denotes an instrument for measuring the
degree of polarization as contrasted to optical rotation. The solution to be measured is first excited in a standard 1-cm cell by
linearly polarized light of appropriate wavelength. The emission
fluorescent beam (with appropriate filters) then passes through a
rapidly rotating polarizer and onto a photomultiplier tube whose
output is fed into an analog computer which calculates the fluorescence polarization, $p = (V - H)/(V + H)$, and the fluorescence intensity, $V + H$. V and H denote relative intensities of vertically polarized and horizontally polarized components in fluorescent light.
Provision is made for automatic deduction of the blank. Polarization
readings can be measured with precision of at least ± 0.002, even at
concentrations of 10^{-11} M fluorescein [5]. Temperature control of
the cell compartment is maintained with an appropriate thermostated
air flow. The response time for the direct readout is about 5 sec.
Mixing times were close to 10 sec. Several other kinetic measurements were made with a stopped-flow polarometer [6,20] whose mixing
and stopping times were less than 5 msec. Reliable experiments in
the millisecond time range could be obtained with fluorescein concentrations as low as 10^{-9} M. The stopped-flow polarometer was
capable of measuring both fluorescence polarization and intensity
simultaneously (Fig. 3B,C,D).

III. EQUATIONS

The ratio of bound to free fluorescent material in Eq. (1) can be directly related to fluorescence polarization and intensity parameters [13]:

$$F_b/F_f = Q_f/Q_b(p - p_f)/(p_b - p) \tag{2}$$

$$F_b/F_f = (Q_f - Q)/(Q - Q_b) \tag{3}$$

Equations (2) and (3) in conjunction with the mass action law in a form derived by Scatchard [21]

$$F_b/F_f = K(F_{bmax} - F_b) \tag{4}$$

can then be used in plots of F_b/F_f vs F_b to determine the equilibrium association constant, K, and F_{bmax}, the number of binding sites available for $\mathfrak{F}$. The concentration of binding sites, F_{bmax}, can be determined by another method which involves examination of titration curves (fluorescence polarization or intensity) which exhibit a sharp break when high concentrations of $\mathfrak{R}$ are titrated with $\mathfrak{F}$ [5]. The ratio of F_b/R at the breaking point is set equal to F_{bmax}. When R is 100-fold larger than the reciprocal of the association constant, F_{bmax} values can usually be computed with an error of less than 5% [5]. Knowledge of the equilibrium constant K as well as the reactant concentrations, F_{fo} and F_{bmax}, provides the investigator with important and necessary guidelines in setting up both classical and relaxation kinetic studies. Such information is mandatory both in utilizing integrated rate methods and in the fitting of reciprocal relaxation times to various equilibrium concentration functions and should in practice be obtained with the same preparations.

One of the simplest and most direct techniques to ascertain the form of an empirical rate law involves the method of initial rates, where only 10 to 15% of the reaction is studied [22]. This laboratory has made extensive use of this classical approach, especially in conjunction with fluorescence polarization and intensity methods

[1 to 4, 6,7,9,16a]. The initial rate of change of polarization which can be readily derived from Eq. (2) is as follows [1]:

$$\left(\frac{dp}{dt}\right)_0 = \frac{Q_b}{Q_f}\,(p_b - p_f)k(R_0)^{N_1}(F_{f0})^{N_2-1} \tag{5}$$

whereas the initial rate of fluorescence intensity change obtained from Eq. (3) is:

$$\left(\frac{dI}{dt}\right)_0 = k(Q_b - Q_f)(R_0)^{N_1}(F_{f0})^{N_2} \tag{6}$$

The order of reaction with respect to R, N_1 can be obtained from the slope of a plot of log (initial rate); i.e., log $(dp/dt)_0$ or log $(dI/dt)_0$ vs log (R_0) at constant F_{f0}, while the order of reaction with respect to F_{f0}, N_2 can be determined by a plot of log (initial rate) vs log (F_{f0}) at constant (R_0). R_0 and F_{f0} refer to the nonfluorescent and fluorescent initial reactant concentrations, respectively. These equations are particularly useful in that the form of the empirical rate law can be determined without a knowledge of the absolute concentration of the reactants. However, once the rate law has been tested over a wide concentration range (factors of at least 5 to 10), the rate constant can be evaluated in R_0 and F_{f0} concentrations are known or can be estimated. It should be noted that if there is a large quenching of the fluorescence intensity of 𝔍 as it attaches to ℜ (Q_f/Q_b is large), appreciable changes in p can occur only when the reaction has proceeded to a great extent. Hence, initial rate conditions no longer hold, p does not approach p_f, and the following equation [see Ref. 16a] must be used instead of Eq. (5):

$$\frac{\dfrac{dp}{dt}\left(\dfrac{p_b - p_f}{(p_b - p)^2}\right)\dfrac{Q_f}{Q_b}}{1 + \dfrac{Q_f}{Q_b}\left(\dfrac{p - p_f}{p_b - p}\right)} = k(R)^{N_1}(F_f)^{N_2-1} \tag{7}$$

R and F_f are no longer at their zero time concentrations but have to be calculated from Eq. (2) or (3) and the value of M. Various inte-

grated rate expressions can also be formulated in terms of fluorescence polarization and intensity parameters by combining classical
integrated rate equations [22] with Eqs. (2) and (3) (Table 1). The
most useful integrated rate expression involves pseudo first-order
conditions, under which the reactant $\Re$ is in such large excess that
its concentration remains virtually constant during the time course
of reaction. The left-hand side of Eq. (1) in Table 1 is usually
plotted against time to obtain the pseudo first-order rate constant
k'. The advantage of using the pseudo first-order rate method is
that (like the initial rate method) a knowledge of the absolute
reactant concentrations is not necessary to determine the form of
the rate law. If the pseudo first-order plot exhibits linearity
over wide ranges of relative concentrations of F_f, the reaction can
be presumed to be first order with respect to F_f. The order of
reaction with respect to R can be determined by plotting log k' vs
R_0 to obtain the slope N_1. In fact, only the concentration of reactant $\Re$ (that in excess of F_f) is necessary to evaluate the magnitude of the overall rate constant. The other integrated rate equations listed in Table 1 are of use only if the absolute reactant
concentrations are known. The simplest expression involves use of
equal concentrations of reactants, whereas the more complicated ones
involve unequal concentrations and/or back reaction. All of these
latter expressions presume a simple second-order rate process for
the forward reaction and a first-order process for the back reaction.
If plots of the left-hand side of these equations vs time are not
linear or do not yield second-order rate constants which are independent of initial concentrations of reactants, the investigator
must resort to differential rate analysis (such as the initial rate
method) to analyze and determine a more complicated rate law. Unfortunately, this has been rarely done. Alternatively, he can turn
to the study of the reaction near equilibrium by means of temperature and/or concentration jump techniques [23,24], which can be
readily adapted to fluorescence intensity [9,25,26] and fluorescence
polarization [16a] measurements. When a process at equilibrium,

such as depicted in Eq. (1), is suddenly slightly perturbed, the
kinetics of re-equilibration can be described by a linear first-
order differential equation [23,24]:

$$\frac{d\Delta F_f}{dt} = -\frac{1}{\tau} \Delta F_f \tag{8}$$

where τ is defined as the relaxation time, and ΔF_f is the deviation
of F_f from its equilibrium value. Since ΔI and Δp are proportional
to ΔF_f near equilibrium, Eq. (8) simply becomes

$$\frac{d\Delta p}{dt} = -\frac{1}{\tau} \Delta p \tag{9}$$

or

$$\frac{d\Delta I}{dt} = -\frac{1}{\tau} \Delta I \tag{10}$$

The reciprocal relaxation time, $1/\tau$, is obtained from the slope of
a plot, $\log \Delta p$ or $\log \Delta I$ vs time, for a given set of reactant equilibrium concentrations. A series of relaxation times, obtained over
a wide range of concentrations, are then fitted to the properly derived expression involving the appropriate equilibrium reactant
concentrations [24]. Different equilibrium expressions are utilized
for different presumed mechanisms. As an example, the equation for
a simple bimolecular association process followed by a unimolecular
dissociation is

$$1/\tau = k_1[(\overline{F_f}) + (\overline{F_{bmax} - F_b})] + k_{-1} \tag{11}$$

It should be pointed out that in order for relaxation data to be of
real value and interpretable in a definitive manner, the form of the
mass law expression must be ascertained unambiguously. This requires
that K and F_{bmax} be obtained over a wide range of concentrations.

IV. RESULTS

A series of typical fluorescence polarization and intensity titration curves obtained for a hapten, antihapten reaction; an enzyme,

TABLE 1

Integrated Rate Expressions Involving Fluorescence Polarization and Intensity Parameters

Expression	Conditions
(1)[a] $\ln\left(\dfrac{I_\infty - I_o}{I_\infty - I}\right) = k't$ where $k' = k_1(NADPH)_o^{N_2}$	Pseudo first-order reaction; NADPH in large excess; no back reaction
(2)[b] $\ln\left(1 + \dfrac{Q_f}{Q_b}\dfrac{p - p_f}{p_b - p}\right) = k''t$ where $k'' = k_1(E_o)^{N_1}$	Pseudo first-order reaction; enzyme in large excess; no back reaction
(3) $\dfrac{Q_f}{Q_b}\dfrac{[(p - p_f)/(p_b - p)]}{A} = k_1t$	Second-order reaction; equal reactant concentrations; no back reaction
(4) $\dfrac{I_o - I}{I - I_\infty} = k_1t$	Second-order reaction; equal reactant concentrations; no back reaction
(5) $\dfrac{1}{A - B}\ln\left[1 + \dfrac{Q_f}{Q_b}\left(\dfrac{p - p_f}{p_b - p}\right)\left(\dfrac{B/A(Q_f/Q_b)[(p - p_f)/(p_b - p)]}{1 + (Q_f/Q_b)[(p - p_f)/(p_b - p)]}\right)\right] = k_1t$	Second-order reaction; unequal reactant concentrations; no back reaction; where A,B = reactant concentration

$$(6)\quad \frac{1}{A - B}\ln\left[\frac{(I_\infty - I_o) - (I - I_o)B/A}{(I_\infty - I)}\right] = k_1 t$$

Second-order reaction; unequal reactant concentrations; no back reaction; where A,B = reactant concentration

$$(7)\quad \frac{X_E}{A^2 - X_E^2}\ln\left[\frac{X_E(A^2 - XX_E)}{A^2(X_E - X)}\right] = k_1 t$$

where $X = A[(I_o - I)/(I_o - I_b)]$

$$X_E = A[(I_o - I_E)/(I_o - I_b)]$$

Equal concentrations; back reaction; A = reactant concentrations

$$(8)\quad \frac{X_E}{A^2 - X_E^2}\ln\left[\frac{X_E(A^2 - XX_E)}{A^2(X_E - X)}\right] = k_1 t$$

where $X = A\dfrac{(Q_f/Q_b)[(p - p_f)/(p_b - p)]}{1 + (Q_f/Q_b)[(p - p_f)/(p_b - p)]}$

$$X_E = \frac{A(Q_f/Q_b)[(p_E - p_f)/(p_b - p_E)]}{(Q_f/Q_b)[(p_E - p_f)/(p_b - p_E)] + 1}$$

Equal concentrations; back reaction; A = reactant concentrations

---

[a] I, I_o, and I_∞, fluorescence intensity at times t, t_o, and t_∞. [b] k_1, second-order rate constant; Q_f/Q_b, ratio of fluorescence intensity of completely free NADPH to completely bound NADPH; p, fluorescence polarization at time t; p of free and bound NADPH are p_f and p_b, respectively. [c] X, amount of initial concentration which has reacted in time t. Subscript E refers to equilibrium value. I_b refers to fluorescence intensity of completely bound NADPH.

substrate reaction; and an antigen, antibody reaction are shown in
Fig. 1. A fixed amount of R (nonfluorescent receptor) is titrated
with various increments of fluorescent or fluorescent-labeled material, and Eq. (2) or (3) was used in conjunction with the value of
M at each point during the titration to calculate appropriate values
of F_b/F_f and F_b. These systems represent situations in which the
fluorescent molecule $\mathfrak{F}$, as a result of reaction with $\mathfrak{R}$, shows quenching (Fig. 1A), enhancement (Fig. 1B), or no change (Fig. 1C), along
with an increase in fluorescence polarization.

Examples of some of the linear Scatchard plots [Eq. (4)] used
to obtain values of K and F_{bmax} from this type of data are shown for
the fluorescein, antifluorescein, and the NADPH, dihydrofolate reductase systems in Fig. 2. A modified version of Eq. (4) involving
a Sips distribution of binding sites [13] has been applied in the
past to antigen, antibody reactions which exhibit heterogeneity.
Such heterogeneity of binding may be indicated by the curved Scatchard plots shown in Fig. 2C. There has also been recent evidence
which indicates that some antibody populations do not involve a Sips
distribution [27]. The magnitudes of the association constants of
several systems which have been studied in our laboratory are listed
in Table 2. It is important to note that whenever both fluorescence
polarization and intensity measurements could be made on a single
system, each method has yielded equilibrium parameters which are
within experimental error of each other.

Several examples of the kinetic curves obtained in studies
where $\mathfrak{R}$ and $\mathfrak{F}$ are mixed and changes in the fluorescence polarization
and/or intensity are monitored during the time course of the reaction
are shown in Fig. 3. As in the equilibrium case, $\mathfrak{F}$ may or may not
exhibit a change in fluorescence intensity during its reaction with
$\mathfrak{R}$ while the polarization increases. Antigen, antibody reactions are
slow enough (k values are ~ 10^5 to 10^6 M^{-1} sec^{-1}) that they can be
studied by classical mixing of dilute reactants in the "static" polarometer, as previously described. Fluorescein-ovalbumin, antiovalbumin (Fig. 3A) and dansyl-BSA, antiBSA represent such antigen,

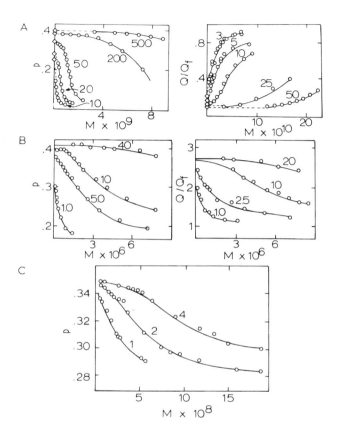

FIG. 1.   Equilibrium fluorescence polarization and intensity
curves involving the titration of fixed amounts of receptor, $\mathfrak{R}$, by
fluorescent ligand, $\mathfrak{F}$. Q/Q_f denotes fluorescence intensity of solution normalized to the fluorescence intensity of the free material.
M refers to total concentration of fluorescent material. A.   Antifluorescein with fluorescein (490-520 nm).   Reprinted from Ref. 5
by courtesy of Academic Press.   B.   L. casei dihydrofolate reductase,
$\mathfrak{R}$, with NADPH, $\mathfrak{F}$ (360-460 nm).   (Reproduced from data in Ref. 9.)
C.   AntiBSA with dansyl-BSA (365-520 nm).   Reprinted from Ref. 14
by courtesy of Pergamon Press.

antibody systems. Hapten, antihapten reactions are somewhat faster

(Fig. 3B) and are probably best studied with the stopped-flow fluo-

rescence polarometer, which can monitor polarization and intensity

data simultaneously in a few milliseconds (Figs. 3B and 3C). How-

ever, if the hapten, antihapten reaction involves such highly fluo-

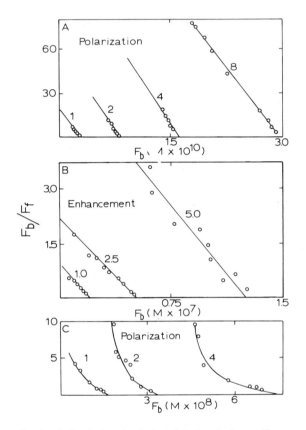

FIG. 2. Typical Scatchard plots obtained from fluorescence polarization and intensity data. A. Fluorescein, antifluorescein
(fluorescence polarization). B. NADPH, dihydrofolate reductase
(fluorescence intensity enhancement). C. Dansyl-BSA, antiBSA (fluorescence polarization). Reproduced in part from Ref. 14 by courtesy
of Pergamon Press.

rescent reactants as fluorescein, the reactants may then be diluted
for appropriate "slow" kinetics (hand mixing [6]).

    Polarization and intensity changes which occur when NADPH and
dihydrofolate reductase react indicate that there are at least two
steps during the binding process. First, there are relatively large
and rapid NADPH fluorescence intensity and polarization increases
(Fig. 3C) which are concentration dependent. A second much slower

TABLE 2

Equilibrium Association Constants for Some
Antigen-Antibody and Enzyme-Substrate Reactions as
Determined by Fluorescence Polarization and/or Intensity[a]

Method	Antibody	Antigen	$K^{(b)}$ (M^{-1})	a	Reference
F.P.	AntiFO	FO	2.1×10^8	0.84	[1]
F.P.	AntiO	FO	1.8×10^8	0.65	[1]
F.P.	AntiBSA	Dansyl-BSA	2.0×10^8	0.61	[1]

Method	Antibody	Hapten			
F.P.	AntiPenRSA	PDABF	9.0×10^6	0.7	[2]
F.P.	AntiFO	F	6.5×10^{10}	1.0	[3]
F.I.	AntiFO	F	6.1×10^{10}	1.0	[3]
F.P.	AntiFO(F_{ab})	F	10^{11}	—	[4]

Method	Enzyme	Substrate			
F.P.	Dihydrofolate reductase	NADPH	5.2×10^6	1.0	[5,6]
F.I.	same	NADPH	5.7×10^6	1.0	[5,6]
F.P.	same	FH_2	2.0×10^6	1.0	[7]
F.I.	same	FH_2	2.0×10^6	1.0	[7]

[a]Abbreviations:

BSA	bovine serum albumin
Dansyl	1-dimethyl aminonaphthalene-5-sulfonyl
F	fluorescein
F_{ab}	univalent antibody obtained by papain degradation
FH_2	dihydrofolate
F.I.	fluorescence intensity
FO	fluorescein-labeled ovalbumin
F.P.	fluorescence polarization
NADPH	reduced form of nicotinamide adenine dinucleotide phosphate
O	ovalbumin
PDABF	fluorescent penicilloyl hapten [2]
Pen	penicilloyl
RSA	rabbit serum albumin

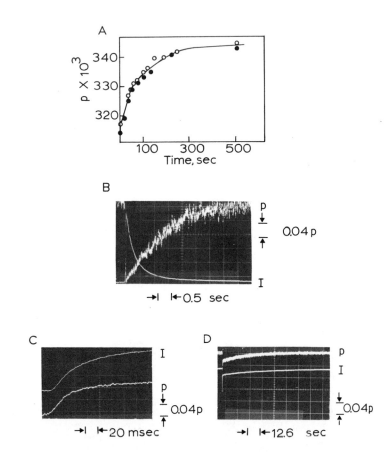

FIG. 3. Fluorescence polarization and intensity kinetic curves involving that rate of association between ℛ and ℑ. A. Fluorescein-ovalbumin, ℑ, antiovalbumin, ℛ. Reprinted from Ref. 1 by courtesy of Pergamon Press. B. Fluorescein, ℑ, antifluorescein, ℛ. C. NADPH, ℑ, dihydrofolate reductase, ℛ. D. Slow secondary fluorescence polarization and intensity changes following the rapid primary reaction between NADPH and dihydrofolate reductase. Fluorescence polarization (upper curve) and intensity (lower curve). C and D reproduced from data in Ref. 9.

step follows (Fig. 3D), occurring in the 5 to 10 sec time range. This slow step involves somewhat smaller polarization and intensity increases and appears to follow a first-order integrated rate law independent of initial reactant concentrations. Some fluorescein, antifluorescein preparations also have revealed a slow first-order

rate step following rapid combination [16a]. Here, fluorescein
initially undergoes rapid large polarization increases and intensity
decreases, followed by slower decreases in both polarization and
intensity (Fig. 3A).

An example of the determination of the form of the empirical
rate law for the primary binding of $\mathfrak{F}$ to $\mathfrak{R}$ is shown for the dihydrofolate reductase, NADPH system in Fig. 4, where the order of reac-
tion with respect to enzyme, N_1, was determined by the method of
initial rates [Eq. (5)]. The value of N_1, as indicated, was determined to be equal to 1 over a wide concentration range in enzyme by
three different methods: increase of the polarization of NADPH,
enhancement of NADPH fluorescence, and quenching of enzyme fluorescence. The order of reaction with respect to NADPH, i.e., N_2, was
also shown to be equal to 1 by these different methods [7,9]. Evaluation of the rate constants for this primary stage was made by em-
ploying appropriate extrapolated values for optical constants Q_b,
P_b, P_f, Q_f, as well as the proper initial concentration of each reactant. As indicated in Table 3, the values of the second-order
rate constants obtained by the different methods were within experimental error of each other. This initial rate approach has also
been utilized to determine empirical rate laws and rate constants
for several antigen, antibody and hapten, antihapten systems (Table
4). Examples of the determination of N_1 and N_2 for an antigen,
antibody reaction are shown in Fig. 5. It should be pointed out
that while only polarization data were used to characterize the antigen, antibody systems, polarization and quenching data were both
utilized in the case of the fluorescein, antifluorescein system.
Furthermore, the polarization rate constant was within experimental
error of the "intensity" rate constant.

As mentioned previously, in addition to initial rate studies,
various integrated rate equations (Table 1) can be utilized to de-
scribe the interaction between $\mathfrak{F}$ and $\mathfrak{R}$ during most of the time course
of reaction. A typical pseudo first-order rate plot involving flu-
orescence intensity changes of fluorescein as it reacts with excess

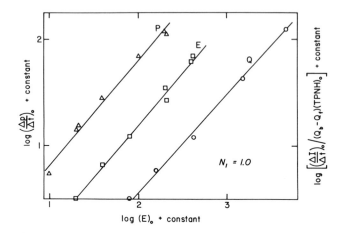

FIG. 4. Determination of N_1, the order of reaction with respect to dihydrofolate reductase by initial rates utilizing three different methods to follow the extent of reaction. $\triangle$, fluorescence polarization (p) of NADPH [log $\Delta p/\Delta t$ vs log $(E)_0$]; $\square$, fluorescence enhancement (E) of NADPH; O, fluorescence quenching (Q) of enzyme. The latter systems involve plots of log $(\Delta I/\Delta t)_0/(Q_b - Q_f)$ vs log $(E)_0$. I denotes fluorescence intensity. Reproduced from data in Ref. 9.

antibody is shown in Fig. 6A, where the left-hand side of Eq. (1) in Table 1 is plotted against time. Similar plots have been obtained which utilize fluorescence intensity changes for the NADPH, dihydrofolate reductase system. For example, dihydrofolate reductase was reacted in large excess of NADPH and the fluorescence quenching of the enzyme monitored (Fig. 6B). These plots were found to be linear throughout the time course as long as R is in large enough excess. The slopes of such plots yield the pseudo first-order rate constant k', which equals $k_1 R_0$. To show that a reaction rate is second order by this approach, one must first demonstrate that k_1' is independent of the initial concentration of F_f and proportional to the initial concentration of R over a wide concentration range. Results implying that the primary reaction between excess NADPH and dihydrofolate reductase is indeed following pseudo first-order kinetics, and an overall reaction order of two are shown in Table 5. Examples of

TABLE 3

Second-Order Rate Constants for Interaction of NADPH with Enzyme[a]

Measurement	k_1		Concentration range			
	Initial rate ($M^{-1} sec^{-1} \times 10^{-6}$)	Integrated rate ($M^{-1} sec^{-1} \times 10^{-6}$)	NADPH ($M \times 10^7$)	Enzyme ($M \times 10^7$)	NADPH ($M \times 10^7$)	Enzyme ($M \times 10^7$)
Fluorescence polarization of NADPH	5.1 ± 0.7[b]	5.8 ± 0.6	1-40	1-20	2-40	2-20
Fluorescence intensity of NADPH	5.1 ± 0.6	5.4 ± 0.6	1-40	1-20	2-40	2-20
Fluorescence intensity of enzyme	4.7 ± 0.7	6.1 ± 0.4[c]	1-20	1-30	10-50	0.3-10

[a] All experiments were performed in 0.005 M Tris, pH 7.5, containing 10^{-5} M EDTA; temp., 18.5°C.

[b] In absence of EDTA, $k_1 = 2.6 \times 10^6$.

[c] Integrated pseudo first-order conditions (NADPH in large excess).

TABLE 4

Some Rate Parameters for Hapten, Antihapten and Antigen, Antibody Reactions

Kinetic method [a]	System [b]	k_1 (M^{-1} sec^{-1})	$k_{-1} = k_1/K$ (sec^{-1})	K (M^{-1})	Ref.
F.P.	Fluorescein, antifluorescein (F_{ab})	8.4×10^7	10^{-3}	10^{11}	[5,6]
F.Q.	Fluorescein, antifluorescein (F_{ab})	8.8×10^7	10^{-3}	10^{11}	[5,6]
F.P.	Fluorescein, antifluorescein (divalent)	6.0×10^7	10^{-3}	1.1×10^{11}	[5,6]
F.Q.	Fluorescein, antifluorescein (divalent)	7.0×10^7	10^{-3}	1.2×10^{11}	[5,6]
F.Q.	DNP-lysine, antiDNP (divalent)	5.0×10^7	1.1	4.5×10^7	[37a]
T.J.	Nitrophenyl, antinitrophenyl (divalent)	1.8×10^8	760	2.4×10^5	[43]
F.P.	Fluorescein-ovalbumin, antiovalbumin (divalent)	4.8×10^5	2×10^{-3}	2.4×10^8	[4]
F.P.	Dansyl-BSA, antiBSA (divalent)	3.4×10^5	2×10^{-3}	1.7×10^8	[4]

[a] F.P., F.Q., and T.J. denote fluorescence polarization, fluorescence quenching, and temperature jump, respectively.

[b] F_{ab}, univalent antibody; DNP, dinitrophenyl; BSA, bovine serum albumin.

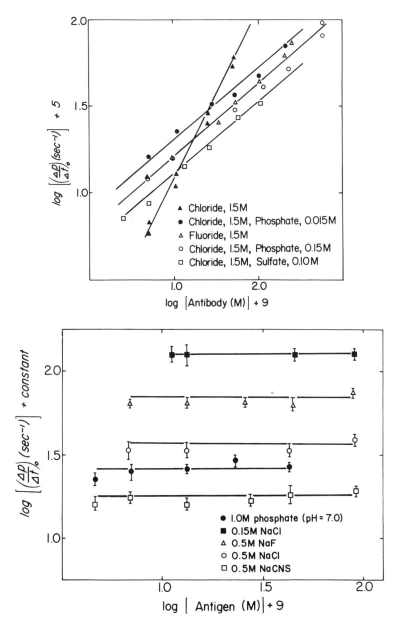

FIG. 5. Effect of specific salts on the order of reaction with respect to antibody N_1 and on the order of reaction with respect to antigen N_2 for some antigen, antibody systems. The empirical rate law involves a fractional order, $N_1 = 0.5$, in media composed of non-chaotropic ions, whereas in all chaotropic media, N_1 and $N_2 = 1.0$. All solutions were buffered at pH 7.0 at 1.5°C. Top, fluorescein-ovalbumin, antiovalbumin. Bottom, dansyl-BSA, antiBSA. Reprinted from Ref. 4 by courtesy of the American Chemical Society.

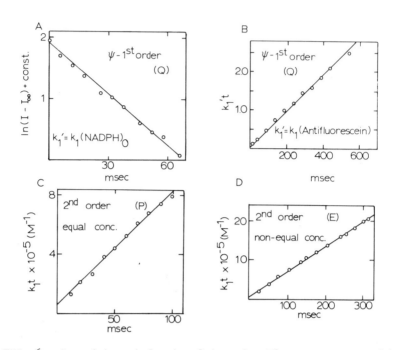

FIG. 6. Some integrated rate plots. A. Fluorescence quenching
of dihydrofolate reductase measured under pseudo first-order conditions. NADPH, in excess, 4.6×10^{-6} M, enzyme (at zero time), 3.5
$\times 10^{-7}$ M. B. Fluorescence intensity of fluorescein measured under
pseudo first-order conditions. Antifluorescein (in excess), $5.0 \times$
10^{-8} M, fluorescein, 5.4×10^{-10} M. k_1 refers to pseudo first-order
rate constant. C. Fluorescence polarization of NADPH under second-
order and equal initial concentrations (2.1×10^{-6} M) of NADPH and
enzyme. D. Fluorescence intensity enhancement under second-order
conditions and unequal concentrations. Enzyme and NADPH at zero time
were 1.0×10^{-6} M and 4×10^{-7} M, respectively. k_1 refers to
second-order rate constant.

other integrated rate plots which involve changes in polarization

or intensity of NADPH are shown in Fig. 6C,D and are linear up to

about 90% completion of the first stage of reaction. The values of

the second-order rate constants have been directly obtained from the

slopes of these integrated rate plots. Several kinetic parameters

are summarized for the dihydrofolate reductase and antifluorescein

systems in Tables 3 and 4. What is most striking is that all the

kinetic methods that we have utilized have yielded second-order rate

TABLE 5

Pseudo First-Order and Second-Order Rate Constants
for Interaction of NADPH with Dihydrofolate Reductase[a]

NADPH ($M \times 10^7$)	Enzyme ($M \times 10^7$)	k_1' [b] (sec^{-1})	k_1 ($M^{-1} sec^{-1} \times 10^{-6}$)
48.5	0.35	31.5 ± 2.4	6.4 ± 0.5
48.5	1.05	31.5 ± 2.4	6.4 ± 0.5
48.5	3.5	27.1 ± 1.2	5.5 ± 0.3
48.5	3.5	28.6 ± 3.4	5.8 ± 0.7
48.5	10.6	30.0 ± 4.1	6.1 ± 1.0
48.5	10.6	27.0 ± 3.7	6.2 ± 0.8
48.5	3.5	27.1 ± 1.2	5.6 ± 0.3
46.0	3.5	27.0 ± 3.2	5.9 ± 0.7
10.6	1.05	6.9 ± 0.5	6.5 ± 0.5
10.6	0.35	6.9 ± 0.5	6.5 ± 0.5

[a]Experiments were carried out in 0.5 M NaCl, 0.005 M Tris, pH
7.5, at 18.5°C in presence of EDTA.

[b]$k_1' = k_1$ (NADPH).

constants for a given system which were within experimental error
of each other. It should be emphasized that these results pertain
to the more rapid primary stage of reaction and not to the slower
process which may follow (Fig. 4). This secondary process found in
some fluorescein, antifluorescein preparations [16a] as well as the
NADPH, dihydrofolate reductase system [9] seems to follow a first-
order integrated rate law whose rate constant is independent of
initial reactant concentration.

Once the form of the empirical rate law has been ascertained
for a given stage of the reaction, e.g., the primary combination,
further characterization of the reaction mechanism can be obtained
by systematic studies of the effects of temperature, pH, and
ionic strength. For example, we have used polarization to show

that the rate of reaction is markedly influenced by the nature of
the ionic medium for several antigen, antibody reactions, namely:
ovalbumin, antiovalbumin; BSA, antiBSA; and fluorescein-IgG, antifluorescein [4]. An interesting relationship was shown to exist
between the Hofmeister series and the empirical rate law. In the
presence of chaotropic ions such as thiocyanate, perchlorate, and
chloride, a simple second-order rate law is followed:

$$- \left(\frac{d(AG)}{dt}\right)_{t \to o} = k(AB_o)(AG_o) \tag{12}$$

whereas in the presence of such nonchaotropic ions as phosphate,
sulfate, and fluoride, a more complicated, fractional order rate law
is followed:

$$- \left(\frac{d(AG)}{dt}\right)_{t \to o} = k(AB_o)^{0.5}(AG_o) \tag{13}$$

Chaotropic ions are defined as those ions in the Hofmeister series
which promote macromolecular unfolding and dissociation, whereas the
nonchaotropic ions are those which promote macromolecular folding
and association [28]. An example of these specific salt effects
upon the order of reaction with respect to antibody, N_1, and to antigen, N_2, is shown in Fig. 5. It has been found, however, in kinetic
studies involving univalent antibody fragments, that simple second-
order kinetics are followed in both chaotropic and nonchaotropic
environments [4]. Further investigations of these specific salt
effects upon k_1 and k_{-1} have been carried out by use of the following pseudo first-order integrated rate expression [1,4]:

$$\log (p_e - p) = \log (p_e - p_f) - \left(\frac{k_1(AB_o) + k_1}{2.3}\right)t \tag{14}$$

where p_e refers to the polarization of the solution at equilibrium.
This pseudo first-order expression is much like expressions (1) and
(2) in Table 1, except that $Q_b/Q_f = 1.0$, and there is a contribution
for the back reaction rate constant k_{-1}. Hence, $k' = k_1(AB_o) + k_{-1}$.
This expression assumes a simple second-order process for the forward reaction and a first-order process for the back reaction. In

excess antibody concentration, plots of log $(p_e - p)$ vs time have
been shown to be linear, at least up to the half-life of polarization change [3], and yield the parameter $[k_1(AB_o) + k_{-1}]/2.3$, denoted by S from the appropriate slope. The deviation from pseudo
first-order behavior after the half-time has been reached may be
indicative of site nonuniformity, site depletion, or possibly
changes in the reaction mechanism. Plots of S, $k_1(AB_o) + k_{-1}$, vs
(AB_o) are shown in Fig. 7 for the fluorescein-labeled ovalbumin,
divalent antiovalbumin system. Similar plots for the dansyl-BSA
antigen and its corresponding divalent and univalent antibody part-
ners have also been obtained [4]. The slope and intercept of this
second-order plot yield k_1 and k_{-1}, respectively. Only systems
which obeyed simple second-order kinetics in their initial rate behavior were studied by this particular integrated rate technique.
It was concluded from such studies that these ions exerted their
effects [especially at higher ionic strength (0.5 to 1.5 M) primarily on k_1 and not on k_{-1}]. A general correlation of the magnitude of
the second-order rate constants for these antigen, antibody systems
exists. In particular, the value of the second-order rate constant
in various ionic media increases according to the following sequence:
$SCN^- < ClO_4^- < Cl^- < F^- < SO_4^{2-}$ < phosphate. Recent kinetic investigations of the reaction between the hapten, fluorescein, and antifluorescein [16a] have shown that the second-order rate constant,
even for this simple hapten, antihapten system, is also markedly
affected by these salts and increases according to the above Hofmeister sequence. These specific salt effects have been explained
in terms of the structural rearrangements that occur between an
antigen or hapten and the corresponding antibody during the primary
combination [4]. The fact that the magnitudes of the second-order
rate constants are particularly sensitive to the nature of the specific anion implies that the reaction in Eq. (1) is not diffusion
controlled but that important rearrangements occur prior to or during
the initial combination. Activation energy determinations for the
ovalbumin, antiovalbumin (12 kcal/mole) [2] and the fluorescein,

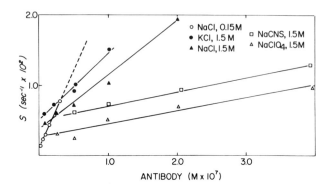

FIG. 7. Effect of different ionic media on integrated rate constants for an antigen, antibody reaction. S denotes the pseudo
first-order parameter, $[k_1(AB_o) + k_{-1}]/2.3$. The slope and intercept
of such plots yield the second-order association constant k_1 and the
first-order dissociation constant k_{-1} for the ovalbumin, antiovalbumin system. All experiments were performed in pH 7.0, 0.01 M Tris
at 1.5°C. Reprinted from Ref. 4 by courtesy of the American Chemical
Society.

antifluorescein (7 kcal/mole) [16a] are well above that expected for
a truly diffusion-limited system, namely, 4 kcal/mole, which results
mainly from the temperature dependence of solvent viscosity.

It has been suggested that divalent antibody clicks open during
its combination with its combining hapten or antigen [29,30]. Such
conformational changes probably involve solvent reorganization and/or
unfolding during formation of the activated complex [2,4]. The idea
of solvent loss is suggested by the fact that estimations of the entropy of activation both for the ovalbumin, antiovalbumin [2] and
the fluorescein, antifluorescein systems [16a] yield much more positive values than those associated with other bimolecular associations
[31]. Furthermore, if the activated complex is less solvated than
the isolated reactants, then anions which compete more effectively
for solvent molecules will tend to promote the reaction rate. Hence,
as observed experimentally, anions with high charge density tend to
promote the reaction rate, whereas anions with low charge density
tend to depress the reaction rate. These solvation effects are also
important in many organic (small molecule) S_N2 reactions in protic

and aprotic solvents [32]. These dehydration effects are important
not only in formation of the activated complex but also in formation
of the final products [33]. This can be concluded from the fact
that similar salt effects on the equilibrium association constants
of hapten, anti-p-azobenzoate hapten [34], and several antigen,
antibody systems [28] have been reported. Furthermore, most antigen,
antibody systems which have been studied by one method or another
during the last several years also appear to have an overall entropy
of reaction which is much more positive than would be expected on
the basis of the loss of translational and rotational motion by the
reactants [35].

     Once the form of the overall mass law and at least some of the
transient kinetic behavior for a particular macromolecular system
have been delineated, temperature-jump techniques provide the logical
method for further characterization of the reaction kinetics near
equilibrium. Typical relaxation traces involving fluorescence intensity changes and fluorescence polarization changes following a
sudden rise in temperature are shown in Fig. 8. Reciprocal relaxation times are plotted as functions of equilibrium concentration of
NADPH and dihydrofolate reductase in Fig. 9. Owing to the sudden
temperature rise and accompanying change in pH (Fig. 8A), an extremely rapid decrease in enzyme fluorescence is noted in the microsecond range. A much slower relaxation time follows. The reciprocal relaxation time $1/\tau$ of this slower equilibrium depends on the
sum of the equilibrium concentrations of enzyme and NADPH at low
concentrations, but approaches a limiting value at high concentrations. This fall-off of $1/\tau$ with concentration implies that at high
reactant concentrations a unimolecular transition may occur which
becomes rate-limiting. A plot of $1/\tau$ vs the sum of equilibrium
concentration NADPH and enzyme can be fitted to the following equation [36]:

$$\frac{1}{\tau} = k_{-2} + \frac{k_2}{1 + k_{-1}/[k_1(\overline{E} + \overline{NADPH})]} \tag{15}$$

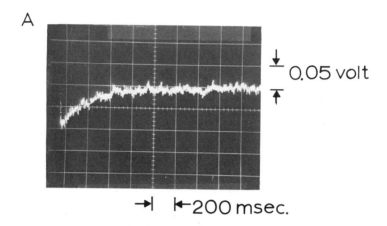

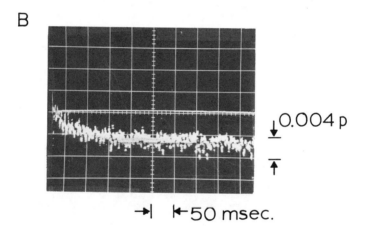

FIG. 8. Temperature-jump fluorescence intensity and polarization data. A. Relaxation fluorescence intensity trace of NADPH reacting with dihydrofolate reductase in 0.005 M Tris, pH 7.5. The temperature of the solution, 18.5°C, was jumped about 5°C. Time scale, 0.2 sec/div; vertical, 0.5 V/div. (Reproduced from Ref. 9.) B. Relaxation fluorescence polarization trace of fluorescent penicilloyl reacting with antipenicilloyl in 0.1 M NaCl, 0.01 M Tris, pH 7.1. Temperature of solution, 17.5°C, was jumped by about 8°C. Time scale, 50 msec/div; vertical, 4 V/div.

A third stage occurs at too slow a rate (5-10 sec time range) to be detected, but it was monitored by stopped-flow measurements (Fig. 3D).

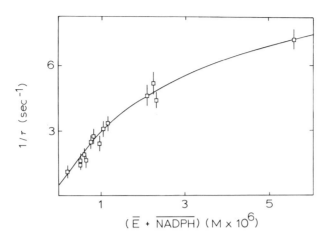

FIG. 9.   Plot of the reciprocal relaxation time, $1/\tau$, vs the sum of the equilibrium concentrations of dihydrofolate reductase and NADPH.   Reactant equilibrium concentrations were calculated from the equilibrium association constant and F_{bmax} values.   Solutions contained 0.5 M NaCl, 0.015 M phosphate, pH 7.5.   K was determined to be 6.0 x 10^6 M^{-1}.   The $1/\tau$ curve was calculated from Eq. (14) using values k_1/k_{-1}, 3.7 x 10^5 M^{-1}, k_2, 7.1 sec^{-1}, and k_{-2}, 0.4 sec^{-1}.   Reproduced from Ref. 9.

Temperature-jump fluorescence polarization studies of the pen-icilloyl, antipenicilloyl system, while still in the preliminary stages, seem to be consistent with a simple bimolecular process be-tween antipenicilloyl and fluorescent penicilloyl hapten [37].

V.   CONCLUDING REMARKS

A general mechanism consistent with the information obtained from the kinetic studies of many of the macromolecular reactions studied in this laboratory [4,9,17] and others [36] would be as follows:

$$\mathcal{R} + \mathcal{F} \underset{k_{-1}}{\overset{k_1}{\rightleftarrows}} \mathcal{R} \ldots \mathcal{F} \underset{k_{-2}}{\overset{k_2}{\rightleftarrows}} \mathcal{R} \ldots \mathcal{F} \underset{k_{-3}}{\overset{k_3}{\rightleftarrows}} \mathcal{R} - \mathcal{F} \tag{16}$$

The initial process would proceed by a relatively rapid bimolecular association followed by somewhat slower intramolecular conversions, which probably involve important structural rearrangements of the reactants during formation of the final complex. The number and nature of the intermediate complexes will be dependent on the particular environmental conditions (e.g., pH, temperature, and ionic media) under which the reaction is taking place. The kinetic behavior of the various intermediate complexes will depend upon whether the reactants are in intimate contact or solvent separated. The primary process, as well as the overall reaction, appears to be driven by a positive entropy, which indicates loss of water and macromolecular unfolding. This solvent release and unfolding could occur as a result of neutralization of charge and/or hydrophobic bonding. The systems we have investigated undergo a primary bimolecular process, which is not diffusion controlled, as evidenced by the effect of specific salts and the magnitude of the measured activation energies. This phenomenon occurs in spite of the fact that the second-order rate constants are quite large (Tables 3,4) and is in contrast to diffusion-limited processes reported for other enzyme, substrate [36] and hapten, antihapten reactions [37a to 39].

Some of what we have designated as conformational changes in the NADPH, dihydrofolate reductase system occur in the 0.1-10 sec time range. While these times are quite slow compared to the millisecond relaxation times reported for other NADH, dehydrogenase complexes [25,26], they are much closer to the time range of conformational changes reported for the NAD^+, glyceraldehyde-3-phosphate dehydrogenase system [40 to 42]. It is surprising, and in a sense novel, that such a small, monomeric enzyme as dihydrofolate reductase would undergo kinetically distinct unimolecular structural changes as NADPH becomes bound. A summary of the kinetic parameters involved in the NADPH-reductase binding is given in Table 6. We have also found in some fluorescein, antifluorescein preparations very slow transitions occurring almost in the minute range [16a]. These structural rearrangements seem to lead to very large equilibrium association

TABLE 6

Kinetic Parameters for Interaction of NADPH with Enzyme[a]

k_1 $(M^{-1} sec^{-1})$	2.6×10^6
k_{-1} (sec^{-1})	7.2
k_2 (sec^{-1})	7.1
k_{-2} (sec^{-1})	0.4
k_3 (sec^{-1})	~0.05
k_{-3} (sec^{-1})	~0.05

[a] Temperature-jump experiments were carried out in 0.5 M NaCl, 0.015 M phosphate, pH 7.5 at 18.5°C.

constants for antigen, antibody and fluorescein hapten, antihapten reactions [4 to 6]. Further work is now being carried out on these fluorescein hapten systems as well as defined antigen reactions with appropriate antibodies. It is anticipated that fluorescence polarization and intensity techniques will continue to provide a sound approach to the detailed kinetic analyses of the more compli- cated enzyme, substrate, and immunological reactions.

Recently we have extended fluorescence polarization equilibrium and kinetic methods to studies involving the interactions of hor- mones with their specific cellular receptors [43 to 45]. The rate of binding between fluorescein-labeled prolactin and membrane mam- mary receptors has been shown to obey second-order kinetics. (Second- order rate constant, $k = 2 \times 10^5$ M^{-1} sec^{-1}.) Preliminary studies of the reaction of fluorescein-labeled estradiol with its cytoplasmic receptor have been initiated utilizing stopped-flow fluorescence polarization. The estradiol appears to undergo a biphasic kinetic binding process at low temperature. From these studies, it becomes quite clear that fluorescence polarization can be extremely useful in the further studies of problems involving hormone-receptor binding.

We have also utilized fluorescence polarization immune assays
to detect both pesticides (diquat, 2-aminobenzimidazole) and hormones such as human growth hormone, insulin, and prolactin. The
fluorescent-labeled compound is reacted with the appropriate antibody in the presence of various amounts of native, unlabeled compound. From the inhibition of the binding equilibrium or inhibition of binding rates [46], hormone or pesticide levels can be
detected in the subnanogram/ml concentration range. The equilibrium association and second-order rate constant for the reaction
between 2-aminobenzimidazole and its antibody was 1.4×10^{10} M^{-1} and
2.5×10^{6} M^{-1} sec^{-1} [47]. Again, fluorescence methods offer great
promise in the detection of organic contaminants in aqueous solution
by immunoassay techniques.

ACKNOWLEDGMENTS

This work was supported by grants from the National Science
Foundation (GB15594) and the National Cancer Institute, National
Institute of Health (A06522).

The author is indebted to A. N. Hicks and R. J. Kelly, and Drs.
W. B. Dandliker, A. J. Portmann, and F. M. Huennekens for their
pleasant and invaluable collaboration.

LIST OF SYMBOLS

Subscripts

e   equilibrium value of parameter

f,b   free and bound forms, respectively,
        of fluorescent material

o   at time approaching zero

(AB)   molar concentration of antibody

(AG)   molar concentration of antigen

(E)   molar concentration of enzyme

$\mathfrak{F}$ symbol for the fluorescent ligand

$\mathfrak{FR}$ symbol for macromolecular complex as defined in Eq. (1)

F_b molar concentration of fluorescent reactant in bound form

F_{bmax} the maximum number of sites available for $\mathfrak{F}$

F_f molar concentration of the fluorescent reactant in free form

I fluorescence intensity, V + H

K equilibrium association constant for the reaction between $\mathfrak{F}$ and $\mathfrak{R}$

k_1 bimolecular association rate constant

k_{-1} unimolecular dissociation rate constant

N_1 order of reaction with respect to $\mathfrak{R}$

N_2 order of reaction with respect to $\mathfrak{F}$

p polarization of fluorescence

Q ratio of fluorescence intensity to F_{fo}

$\mathfrak{R}$ symbol for nonfluorescent receptor

R molar concentration of $\mathfrak{R}$

V,H relative intensities of vertically polarized and horizontally polarized components in fluorescent light

M molar concentration of fluorescent material

REFERENCES

1. W. B. Dandliker and S. A. Levison, _Immunochemistry_, 5. 171 (1967).

2. S. A. Levison, A. N. Jansci, and W. B. Dandliker, _Biochem. Biophys. Res. Commun._, 33, 942 (1968).

3. S. A. Levison and W. B. Dandliker, _Immunochemistry_, 6, 253 (1969).

4. S. A. Levison, F. Kierszenbaum, and W. B. Dandliker, _Biochemistry_, 9, 322 (1970).

5. A. J. Portmann, S. A. Levison, and W. B. Dandliker, _Biochem. Biophys. Res. Commun._, 43, 207 (1971).

5a. A. J. Portmann, S. A. Levison, and W. B. Dandliker, _Immunochemistry_ (1975).

6. S. A. Levison, A. J. Portmann, F. Kierszenbaum, and W. B. Dandliker, Biochem. Biophys. Res. Commun., 43, 258 (1971).

7. S. A. Levison, R. B. Dunlap, L. E. Gundersen, and W. B. Dandliker, Fed. Proc., 30, 872 (1971).

8. F. M. Huennekens, R. B. Dunlap, J. H. Freisheim, L. E. Gundersen, N. G. L. Harding, S. A. Levison, and G. P. Mell, Ann. N. Y. Acad. Sci., 186, 85 (1971).

9. S. A. Levison, L. E. Gundersen, R. B. Dunlap, F. Otting, A. N. Hicks, and F. M. Huennekens, in preparation.

10. D. J. R. Laurence, Biochem. J., 51, 168 (1952).

11. R. F. Steiner, Arch. Biochem. Biophys., 46, 291 (1953).

12. W. B. Dandliker and G. Feigen, Biochem. Biophys. Res. Commun., 5, 299 (1961).

13. W. B. Dandliker, H. C. Schapiro, J. W. Meduski, R. Alonso, G. A. Feigen, and J. R. Hamrick, Jr., Immunochemistry, 1, 165 (1964).

14. F. Kierszenbaum, J. Dandliker, and W. B. Dandliker, Immunochemistry, 6, 125 (1969).

15. W. B. Dandliker, S. P. Halbert, M. C. Florin, R. Alonso, and H. C. Schapiro, J. Exp. Med., 122, 1029 (1965).

16. W. B. Dandliker, R. J. Kelley, J. Dandliker, J. Farquhar, and J. Levin, Immunochemistry, 10, 219 (1973).

16a. S. A. Levison, A. N. Hicks, A. J. Portmann, and W. B. Dandliker, Biochemistry (1975).

17. W. B. Dandliker, in Methods in Immunology and Immunochemistry (C. A. Williams and M. W. Chase, eds.), Vol. III, Academic Press, New York, 1971.

18. W. B. Dandliker and A. J. Portmann in Fluorescence of Macromolecules (R. F. Steiner and I. Weinryb, eds.), Plenum Press, New York, 1971.

19. J. U. White, D. E. Williamson, S. A. Levison, and W. B. Dandliker, in preparation.

20. S. A. Levison, R. J. Kelly, A. N. Hicks, and W. B. Dandliker, in preparation.

21. G. Scatchard, Ann. N. Y. Acad. Sci., 51, 660 (1949).

22. A. A. Frost and R. G. Pearson, Kinetics and Mechanism, John Wiley, Inc., New York, 1961, p. 186.

23. M. Eigen, Disc. Faraday Soc., 17, 194 (1954).

24. M. Eigen and L. de Maeyer, Tech. Org. Chem., 8, Part II, 895 (1963).

25. G. H. Czerlinski and G. Schreck, Biochemistry, 3, 89 (1964).

26. E. J. del Rosario and G. G. Hammes, Biochemistry, 10, 716 (1971).

27. T. P. Werblin and G. W. Siskind, Immunochemistry, 9, 987 (1972).

28. W. B. Dandliker, R. Alonso, V. A. deSaussure, F. Kierszenbaum, S. A. Levison, and H. C. Schapiro, Biochemistry, 6, 1460 (1967).

29. A. Feinstein and A. J. Rowe, Nature, 205, 147 (1965).

30. K. Valentine and N. Green, J. Molec. Biol., 27, 615 (1967).

31. K. J. Laidler, Chemical Kinetics, McGraw-Hill, Inc., New York, 1965, p. 198.

32. A. J. Parker, Advan. Phys. Org. Chem., 5, 173 (1967).

33. F. Haurowitz, Biol. Rev. Cambridge Phil. Soc., 27, 247 (1952).

34. D. Pressman, A. Nisonoff, and G. Radzimski, J. Immunol., 86, 35 (1961).

35. W. C. Boyd, Fundamentals of Immunology, 3rd ed., Interscience, New York, 1956, p. 285.

36. G. G. Hammes and P. R. Schimmel, The Enzymes, 3rd ed., Vol. 2, 1970, p. 67.

37. S. A. Levison and W. B. Dandliker, unpublished results.

37a. L. A. Day, J. M. Sturtevant, and S. J. Singer, Ann. N. Y. Acad. Sci., 103, 611 (1963).

38. A. Froese and A. H. Sehon, Immunochemistry, 2, 135 (1965).

39. A. Froese, Immunochemistry, 5, 253 (1968).

40. B. Chance and J. M. Park, J. Biol. Chem., 242, 5093 (1967).

41. K. Kirschner, J. Molec. Biol., 58, 51 (1971).

42. K. Kirschner, E. Gallago, I. Schuster, and D. Goodall, J. Molec. Biol., 58, 29 (1971).

43. W. B. Dandliker, W. P. Vanderlaan, and S. A. Levison, Pacific Slope Conference, 1974.

44. S. A. Levison, W. P. Vanderlaan, and W. B. Dandliker, Fed. Proc., 34, 584 (1975).

45. W. B. Dandliker, S. A. Levison, and W. P. Vanderlaan, in preparation.

46. D. Murayama, H. R. Lukens, C. B. Williams, A. N. Hicks, S. A. Levison, and W. B. Dandliker, in preparation.